Offshoring Information Technology
Sourcing and Outsourcing to a Global Workforce

The decision to source software development to an overseas firm (offshoring) is frequently looked at in simple economic terms – it is cheaper. In practice, however, offshoring is fraught with difficulties. As well as the considerable challenge of controlling projects at a distance, there are differences in culture, language, business methods, politics, and many other issues to contend with. Nevertheless, as many firms have discovered, the benefits of getting it right are too great to ignore. This book explains everything you need to know to put offshoring into practice, avoid the pitfalls, develop effective offshore strategies and effective working relationships. Split into three parts: offshoring fundamentals; management competencies; and a section on broader issues including a unique look at the viewpoint of an outsourcing provider. Written for CTOs, CIOs, consultants and other IT executives, this book is also an excellent introduction to outsourcing for business and MIS students.

Erran Carmel is an Associate Professor at the Kogod School of Business at American University in Washington DC, USA.

Paul Tjia is a Senior Consultant and founder of GPI Consultancy in Rotterdam, The Netherlands.

Offshoring Information Technology

Sourcing and Outsourcing to a Global Workforce

Erran Carmel

Kogod School of Business, American University, USA

Paul Tjia

GPI Consultancy, The Netherlands

CAMBRIDGE UNIVERSITY PRESS
Cambridge, New York, Melbourne, Madrid, Cape Town, Singapore, São Paulo

Cambridge University Press
The Edinburgh Building, Cambridge CB2 8RU, UK

Published in the United States of America by Cambridge University Press, New York

www.cambridge.org
Information on this title: www.cambridge.org/9780521843553

First published 2005
Third printing 2007

Printed in the United Kingdom at the University Press, Cambridge

A catalog record for this publication is available from the British Library

Library of Congress Cataloging in Publication data

ISBN-13   978-0-521-84355-3 hardback

Contents

List of contributors

Principal authors

All chapters and cases were written by Carmel and Tjia unless indicated otherwise.

Erran Carmel is an Associate Professor at American University's Kogod School of Business in Washington D.C. He co-founded the school's program in Management of Global IT. He is the author of the successful 1999 book *Global Software Teams: Collaborating Across Borders and Time Zones*. He has also written over 70 articles, reports and manuscripts. He consults to a variety of organizations on global software development and is often asked to speak at industry and professional groups around the world on this subject. He has been visiting professor at University College Dublin (Ireland) and Haifa University (Israel).

Paul Tjia is a Senior Consultant and founder of GPI Consultancy in Rotterdam, The Netherlands. He has a background in information technology (IT) and cultural anthropology. He has assisted many clients embarking on the offshore journey with research, feasibility studies, country and partner selection, and due diligence. He also conducts intercultural training and promotes offshoring by organizing seminars and study tours. For offshore providers, he arranges workshops on the marketing of offshore services. He writes articles and reports, and often speaks at offshore industry events. He can be contacted at info@gpic.nl

Biographies of contributors

Kaladhar Bapu is the Manager of Usability Engineering Group at Cordys R&D, India, Asia's best equipped usability laboratory. Prior to Cordys, while at Baan, he was one of the very first to bring usability to India. As a visiting faculty he has been spreading usability awareness among design and technology students at institutes like the Industrial Design Center, IIT Bombay, and IIIT Hyderabad. He founded "usabilitymatters.org", a non-profit organization to make the common man more sensitive towards usability and design. He co-authored the case about offshoring usability that appears in Chapter 9.

Erik Beulen, PhD, is an International Business Development Manager with Atos Origin based in The Netherlands. He is also affiliated with Tilburg University as an Assistant Professor. His papers have been published in academic journals including the *Communications of the AIS* and the *Proceedings of HICSS* and *ICIS Conferences*. He is the author and co-author of various book chapters on outsourcing. He co-authored Chapter 7, Managing the Offshore Transition.

Rebecca S. Eisner is a Partner in the International Law Firm of Mayer, Brown, Rowe & Maw. Her practice focuses on complex global and offshore technology and business process outsourcing transactions. She also advised clients in strategic alliances, joint ventures, licensing, technology development, communications agreements, Internet commerce, data privacy and data transfer issues. She is a frequent writer and speaker on outsourcing, licensing and e-commerce topics. She is also recognized by *Chambers Global – The World's Leading Lawyers* in the area of IT, Communications and Out-sourcing (2003–2004). She wrote Chapter 6, Offshore Legal Issues.

Julia Kotlarsky is a Lecturer in Information Systems at Warwick Business School, UK. She is completing a PhD degree in Information Systems at Rotterdam School of Management, The Netherlands. Julia grew up in Russia and holds an MSc degree in Industrial Engineering and Management at the Technion, Israel. Her research interests revolve around social and technical aspects of globally distributed software develop-ment: coordination, knowledge sharing, social networks, and e-collaborative technolo-gies; as well as component-based design and innovation in software development. She wrote the case about Russian cultural differences that appears in Chapter 9.

Subramanian Ramanathan has been involved in offshore and distributed software development for over 17 years. As General Manager and Director of Baan-India, he was responsible for setting up and operating Baan's product development center in India with over 800 software engineers between 1987 and 2000. He later established the development center for Vanenburg group in India (now called Cordys R&D). He is currently the Managing Director of Vanenburg IT Park, Hyderabad, India. He is also associated with the Hyderabad Software Exporters Association. He co-authored the case about offshoring usability that appears in Chapter 9.

Lu Ellen Schafer is the Executive Director of Global Savvy, based in California. Global Savvy is a consultancy that works with international companies to bridge the cultural gaps between clients and outsourcing partners in India, China and other locations around the globe. It has trained over 11,000 individuals to date from a range of the largest technology companies including Cisco, HP, IBM, Oracle, Accenture, Infosys, Wipro, and HCL. Global Savvy's work has been written about in the

New York Times and the Economic Times of India. Lu Ellen is a frequent speaker at international outsourcing conferences. She wrote the case about cultural miscommunication that appears in Chapter 9.

Peter Schumacher is the founder, President, and CEO of the Value Leadership Group, based in Germany. Peter is an internationalist, having lived and worked in Europe, the USA, and Asia. Early in his career he built a logistics business in Japan and Asia from the ground up. He later worked at Perot Systems in the USA, London, and Munich. He has managed several mergers and acquisitions projects and was the CEO of a European leasing business. Peter has been involved with offshore strategy and operational issues since 1999. He holds an MBA from New York University. He co-authored Chapter 5, Offshore Strategy.

Johan Versendaal is an Assistant Professor at Utrecht University, Institute of Information and Computing Sciences, in The Netherlands. He is the co-architect of the university's newly developed Business Informatics program. His research interests include business–IT alignment, product software development, and human factors and organizational issues of (e-)procurement. Prior to joining Utrecht University, he was the Product Manager at Baan for the purchasing applications, and manager of Baan's usability team. He also worked as a business and usability consultant for Atos Origin. Johan holds a PhD from Delft University of Technology. He co-authored the case about offshoring usability that appears in Chapter 9.

Foreword

Offshoring Information Technology is an appealing book. Appealing not only because it deals with a topic of growing contemporary significance but also because it does so with lucidity, comprehensiveness, thoughtfulness and insightfulness. Over the last decade, offshoring of IT has become a mainstream business phenomenon and, as a result, managing offshoring has emerged as a critical business competence for firms. Erran Carmel and Paul Tjia in this book present a comprehensive treatment of IT offshoring and discuss the competencies required to successfully manage it. Dexterously guiding the readers through the offshore IT landscape and navigating through a range of pertinent topics, this book presents a well-crafted body of knowledge and guidelines to succeed with offshoring of IT. Recounting my experiences over the last 9 years or so, my immediate reaction to this book was: "Why the hell was this book not available a few years ago?" Had it been available, I thought, it would have positively influenced productivity and performance in offshore IT work – and saved sleepless nights for many people!

Circa 1995: The phenomenon of offshoring was starting to gain prominence. Attracted by its low cost structure and the ability to access a global resource pool, many multinational companies had begun leveraging the benefits of offshore IT either through their own subsidiaries or from third-party suppliers. Lured by its promise, I made the transition into the growing IT industry in 1996 to join a subsidiary of Siemens in Bangalore, engaged in communications software development. My job there was to help improve the performance of projects and the resultant quality of software – aspects crucial to establish credibility of an offshore IT organization. Being new to offshore work with no aid available to rely on, I was obviously overwhelmed by the complexities and challenges of the globally distributed work. I struggled hard to successfully deliver on my objectives amidst cultural and time zone differences, geographical separations, and diverse stakeholder expectations.

However, when I took up a new position in 1997 as a member of the management team that was responsible for setting up Lucent Technologies product software R&D center in Bangalore, I received my first full-blown exposure to the world of offshore development and its various nuances. The challenge at hand was to establish a best-in-class offshore software development organization while beating the barriers of time and distance and simultaneously balancing the various considerations (for instance, economic, technical, legal, cultural, organizational and managerial) involved in offshore work. As a general manager, my job also required me to engage with third-party service providers. There were times when the work I managed spanned seven countries!

With no account of proven practices available, I was forced to meet complexities head-on, devising my own ways and learning through perpetual refinement the art of managing offshore work.

Circa 2005: Offshoring of IT is now an irreversible trend and is regarded as a business necessity. Companies across the globe are capitalizing on offshoring to achieve business competitiveness. In the last 3 years or so, offshore IT has assumed new forms to include offshoring of business processes and even R&D. Offshoring of IT is intensifying and firms are strategically leveraging offshore capability and the structural cost savings, while also focusing on deriving operational innovation. Robust models and quality and project management processes are employed to unleash the benefits of offshore IT, such as the *Global Delivery Model* of Infosys Technologies – the company where I currently work as an associate vice president. However, the same complexities and challenges still exist, some even growing in their magnitude and assuming new dimensions. Although many refined and proven managerial and organizational practices, and technological tools and infrastructures, are now available, the challenges and constraints involved in managing offshore IT are far from gone. The art of managing offshore IT work is still evolving.

I consider *Offshoring Information Technology* an important book in many ways. First of all, IT offshoring is part of the larger phenomenon of globally distributed work and while much is understood about globalization of work in general, the body of knowledge on IT offshoring is rather scarce. In this book, Carmel and Tjia provide a structured understanding of the phenomenon of IT offshoring, discuss its various nuances and offer effective practices to succeed with offshoring of IT. As a practitioner-scholar, I have been researching globalization of R&D and software development for about five years now and I am impressed with the systematic and pragmatic coverage of offshore IT Carmel and Tjia have crafted. To the best of my knowledge, this is also the first comprehensive source of knowledge on IT offshoring.

Secondly and very importantly, this book provides practical insights and guidance for managers to help them acquire or refine the competencies required for effectively leveraging IT offshoring. Even though my stints in different organizations in various capacities and settings have allowed me to gain some experience in managing offshore IT, I find this book containing pearls of wisdom. Carmel and Tjia discuss a range of important topics for embarking on and managing offshore IT work. Among other things, this book addresses economics and risks of IT offshoring, assessment and planning for offshoring IT, offshoring strategy, and transition management. It also offers advice on how to alleviate the issues arising out of distance, time zone differences, and cultural diversity in addition to discussing some typical contractual and legal considerations. Notably, the book also presents national policy-level implications for capitalizing on the offshoring wave in addition to offering perspectives on marketing of offshore IT services.

Interestingly, there is also a chapter devoted to discussing the political dimension associated with offshoring.

Both as an executive operating in the midst of accelerating pace of offshoring and as a practitioner-scholar deeply interested in the area of globally distributed work, I believe this book greatly enhances our understanding of a jigsaw puzzle called IT offshoring and equips us well to deal with it. In recording my appreciation for this valuable book, I am also quite hopeful that it will significantly illuminate the people engaged in the business of offshoring IT.

Bangalore, India   Deependra Moitra
21 January 2005   Associate Vice President
                                                      Infosys Technologies Limited

Preface

Why we wrote this book

Whether one is for it or afraid of it, we are convinced that managing offshoring is a competency that tomorrow's IT managers must learn. We wrote this book to help build that competency.

This book builds offshore competency in the breadth and depth of the material covered here: offshore economics, offshore strategy, offshore legal issues, how to get started in offshoring, and many other critical topics.

By teaming up across the Atlantic (Erran is in the United States, Paul is in The Netherlands), we bring different views and challenged each other's assumptions. We bring different views in other ways, we formed a business-academic alliance (Erran is a professor of business and Paul is a consultant on offshoring). We also invited other experts to contribute: there are eight additional authors and co-authors of some chapters and some cases. For example, we invited an attorney specializing in offshoring, Rebecca Eisner, to author the chapter on offshore legal issues.

We have also collected many real-life examples: nine in-depth cases, all of which are first published here, as well as countless stories and anecdotes sprinkled throughout the book.

This book is also a resource for students and teachers in business and technology programs. As we wrote this book the first offshore outsourcing classes were offered. Today, the topic of offshoring should be a component of any management curriculum.

Finally, we also wrote this book for policy makers and analysts in or around governments. Governments in dozens of nations have been devoting more attention to offshoring as path to increase their national wealth.

We, the authors, both live in countries where offshoring has become a controversial political topic. Thus, as we wrote this book, we were often asked by friends and colleagues: "So, what stance are you taking on this issue?" By this our interrogators meant: are we for or against offshoring? In this book we cover the advantages and disadvantages of offshoring openly and honestly. We did not write this book to take a political stance; this is a management book.

Offshore jargon and the scope of this book

Why isn't this book called *offshore information technology OUTsourcing?*

The term *offshore information technology outsourcing* is replete with misleading usage. So, at the outset, we will define and parse some offshore jargon and explain what "they" mean – and what we mean in this book. This will also be a good place to explain what is in the scope of this book.

What is meant by offshore?

Strictly speaking, offshore can be any country outside the home country, similar to the word "foreign." Before everyone began using *offshore IT outsourcing*, the common usage of offshore in the business context was for offshore tax havens,[1] often on small islands offshore, such as the Cayman Islands off the coast of the US. Indeed, an Internet search will still present these items on occasion.

But, the word "offshore" has taken a new meaning. It is understood by many of its business users to mean the shifting of tasks to *low-cost* nations, rather than to any destination outside the country. Low-cost nations are those that fall into the economic grouping of "developing nations" or "emerging nations." Thus a British software firm does not usually refer to its US software research center as an "offshore site." Really, the broader theme of this book is the ascendancy of nations outside the most developed industrialized economies – and the true globalization of the software industry.

"Offshore" has spawned many derivative terms, the most important of which is the opposite: **onshore**. In this usage "onshore services" are those that are provided by foreign firms locally (onsite) often using lower-wage foreign employees. For example, the US special work visa, the H1-B, has been used to import labor in order to staff these "onshore" services. Amusingly, *offshore* has morphed in the hands of marketing departments as the list of terms in Exhibit 1 demonstrates.

What is meant by outsourcing?

Outsourcing has two implications. First, it means that tasks and processes are contracted to be performed outside the boundaries of the firm. Thus, some of GE's offshore development centers (ODCs) in India are, indeed, outsourcing, because they are performed by a third party, Tata Consultancy Services; while Siemens' software development center in India is owned by Siemens and its employees are Siemens' employees.

Many technology firms have globalized via acquisitions – acquiring smaller software firms around the world – and then molding them into their global operations. Other firms have expanded offshore by setting up greenfield subsidiaries – setting up a new, from-the-ground-up subsidiary or software center. When such an offshore center

- Onshore
- Offshore
- Nearshore
- Best Shore (EDS)
- Anyshore (BearingPoint)
- Rightshore (Capgemini)
- Farshore (CG&Y)
- Dualshore (NIIT)
- Offsourcing (HCL Technologies)
- Offshoring
- Nearsourcing
- Nearshoring
- Multishore

Exhibit 1 A collection of marketing-oriented terms for offshore sourcing[2] (the source of the term is noted in parentheses where it is known).

is owned by the client company, then in offshore-speak it is called a **captive center**. There are also hybrid collaborative arrangements, such as setting up a joint venture with a local partner.

So, really, a better term to use, instead of outsourcing, is **sourcing**. This book is about *offshore sourcing*. Where sourcing can be from outside the firm or inside the firm: whether it be outsourcing, or inside the company in captive centers.

Second, the traditional outsourcing industry sees outsourcing more narrowly: when an entire process is delegated to an outsider – a call center, network management, or application support – and sometimes where assets and staff are actually transferred to the outsourcing firm. But, these days, many offshore activities are one-off, single projects that are contracted on a one-by-one basis. Therefore, strictly speaking, this is not outsourcing in the traditional sense, but "project contracting," or **out-tasking**. While we use the term out-tasking in this book, we do not subscribe to the narrow definition of outsourcing.

What is meant by Information Technology – IT?

Some software engineers hear it as *information systems* type activities that are conducted across industries, by end-user organizations, such as a retail chain. We do not segregate IT from software. This book is about *any type of software*-related activity: IT services and IT applications, software products, and embedded software.

Figure 1 has a small appendage hanging from its right side. This is **IT-enabled services (ITES)**. IT-enabled services includes the many services that are now being sliced away and offshored: from call centers, to medical transcription, architectural drafting, through financial securities research. These are not software activities. Nevertheless,

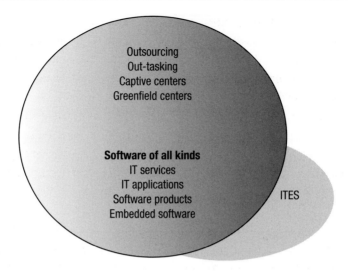

Figure 1 Scope of this book.

IT-enabled services offshoring and IT offshoring are closely tied together. Therefore, we mention IT-enabled services throughout the book, but in particular in Chapters 1 and 10.

The book roadmap

We structured this book so that it does not have to be read linearly. Skim, jump, or hunt for the chapters that are most pressing to you.

Part I, *The Fundamentals* covers the most important issues to the manager, especially in early offshore stages. Chapter 1, *The Offshore Landscape*, gives the reader a broad overview of offshoring past, present, and future, while introducing some of the topics that will be covered in later chapters. Chapter 2, *Offshore Economics and Offshore Risks*, examines the most critical business issue: Is there really a cost advantage? Or is this, perhaps, an illusion? It also includes the first of our nine practical cases: how a giant American company calculates the real costs of offshoring. This chapter also takes a close look at the other side of cost savings: the risks in offshoring.

Chapter 3, *Beginning the Offshore Journey* is written for the manager who, as the title suggests, is just beginning. It deals with the three major phases: laying a solid foundation, the identification of potential service providers, and then selecting the best one. Chapter 4, *The Offshore Country Menu*, gives the reader a foundation for understanding the many countries that are offshore destinations. Even if you are convinced ahead of time that you will offshore to India, this chapter will be useful. The chapter ends with small briefings on a cross section of 11 offshore destinations.

Part II of the book is titled *Managerial Competency*. It takes the business reader through five building blocks of managing offshore activities. The chapters are: *Offshore Strategy*, on the cost strategy and beyond; *Offshore Legal Issues*, covering the contractual concerns and legal risks; *Managing the Offshore Transition*, covering the three difficult topics of knowledge transfer, change management, and governance; *Overcoming Distance and Time*, offering the many small formal and informal solutions to this difficult problem; and *Dealing with Cross-Cultural Issues*, which takes a light-hearted and practical perspective to differences between people around the world.

Finally, in Part III, *Other Stakeholders*, we introduce perspectives of interest to different readers. First, *Building Software Industries in Developing Countries* takes the view of policy makers interested in how their countries can gain from the growing global demand for offshore services. Then comes *Marketing of Offshore Services – the Provider Perspective*, which presents the view of offshore providers' marketing and sales staff seeking to enter new markets and target new clients. Finally, the last chapter examines the controversial political and social implications of offshoring.

Acknowledgments

Marty McCaffrey was the spiritual father of this book in two important ways. First, Marty began discussing the need to write a book on offshoring as far back as 2000. Second, Marty made many professional introductions to us – many of which benefited this book directly or indirectly.

We received gracious support on several of the cases. We wish to thank: for the Intel case, Eleanor Wynn, Cynthia Pickering, Tammy Hertel, and Nathan Zeldes; and for the GroupSystems case, Bob Briggs. Several others provided us generous access to write our cases. These are the people behind the anonymous, but true, cases in Chapters 2, 4, and 7. We wish we could thank them by name because their contribution was significant, but unfortunately we cannot because they requested to remain anonymous.

The following colleagues kindly read draft chapters and offered excellent improvements: Frank DuBois, Alberto Espinosa, Sally Fowler, Zerubbabel Johnson, Jennifer Oetzel, Steve Sawyer, and Jeremy Wells. Others helped in direct and indirect ways: Bill McHenry, Brian Nicholson, Eric Olsson, V. Sridhar, and Shirley Tessler.

Erran adds: My most committed reader of many drafts and my most unrelenting critic has been my father, Eli, a global business manager and also a former professor. He persisted in demanding more and more. My thinking benefited immensely from his comments and it was rewarding to work together.

Paul adds: Dedicated to the memory of my father, Tian Seng Tjia. His business advice and moral support were of enormous value when starting an offshore consultancy in the middle of the 1990s. His motto "keep on fighting" also proved valuable on several occasions when writing this book.

Part I

The fundamentals

1 The offshore landscape

Offshoring IT work is an important milestone in the history of global economics. Why has this happened now?

There is no *one* factor that brought about this phenomenon, but rather, six. Six principal forces converged, as depicted in Figure 1.1. The first of these forces is well known: the *globalization* of trade and, more recently, the globalization of trade in services, which is now approaching 2 trillion USD annually. Borders began opening in the 1980s as market-based solutions gained broad acceptance. The collapse of the Soviet bloc spurred this process even more.

Nations that were once hostile to business, or at best indifferent, are now competing with one another to attract foreign investment and spur their software sectors, creating a *business-friendly climate*. Nations are offering tax incentives and are easing government regulations. They are building technology parks to make it easy to set up and run business

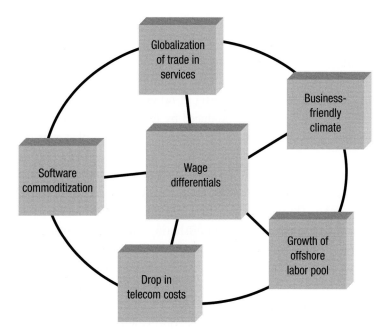

Figure 1.1 The principal economic, business, and technology forces of offshoring.

operations. India and the Philippines are dotted with such technology parks. China has established 15 software parks and 53 technology parks (more on the country "menu" will come in Chapter 4).

Meanwhile, the number of engineers pouring out of universities and technical schools in India, China, and other nations has surged. China, alone, graduates four times as many engineers as the US every year. While the quality of these programs was once inferior to those in industrialized nations, the gap has narrowed. The elite of the offshore labor pool – the talent that is now being directed at higher-end software activities (e.g. research and development (R&D)) – was always there. But, not long ago, this talent would emigrate to the industrialized nations or find other jobs. Today, global technology firms tap these talented engineers and scientists wherever they may be.

In the course of just a decade *communications costs* have decreased to almost zero for nearly unlimited usage. This has brought about a remarkable outcome – that it is almost as easy to work with someone across the ocean as across town (though not equally the same, see Chapters 8 and 9). Between the late 1990s and the early 2000s the benchmark international calling rates have fallen by 80–90%; that is, for those who still use standard rates. Many software workers use voice over IP at zero marginal cost. Equally important for software, the bandwidth has expanded by orders of magnitudes, from almost zero in the 1990s, reaching 4 gigabits per second to India alone in 2004 (with up to 9 terabits per second of system capacity). It was only in 1994 that one of the pioneering project managers offshoring to India had his team copy the weekly software "build" onto tape every Friday just in time for the FedEx pick-up that would fly the tape across the ocean.

Software commoditization is not as well understood by those outside the software industry. It is the standardization of software development practices and tools. For the first time, in software's 50-year history, some software tasks are sufficiently routinized and automated that they have been "commoditized." These tasks are nearly undifferentiated by producer, like a barrel of oil or a bushel of wheat. Once some tasks are commoditized they can be produced by the lowest-cost, most-productive bidder. As one manager commented to us "these are the skills that you can shop for on the Internet."

Finally, and make no mistake about it, the dominant force in offshoring is the *wage differential* between low- and high-wage nations. Hence this force appears in the center of Figure 1.1. The wage differentials lead to lower costs. Some managers will utter other politically acceptable reasons for offshoring that seem less offensive than simply slashing costs, but these are often secondary considerations voiced for appearances. The cost pressures have made offshoring a strategic necessity for some firms (see Chapter 5, Offshore Strategy). Not only are corporate executives stressing cost savings, but American venture capital firms, in an effort to reduce their own capital investments in young firms, have pushed technology startups to perform their R&D offshore from the outset. Until the 1990s, technology firms looked largely at labor pools in high- and middle-wage nations: the G7 nations,[1] Switzerland, Israel, Brazil, and several others. This has permanently changed.

To be sure, there are other, secondary, forces that helped spur offshoring, such as the emergence of sophisticated IT firms offshore, especially in India; and the advantages, in isolated cases, of working around the clock. Additionally, market access has been a factor for large technology companies. Large global firms need to, or are forced to, invest in operations in important nations. China is the premier example of this. No important technology firm can sell to China today without having some R&D or manufacturing operations in-country. For example, in 2003 China mandated its own cryptographic standard for wireless local area networks (LANs). Foreign firms who wanted to access this market were forced to collaborate in software R&D with local companies.

Historical context and lessons for the future

Offshoring is not new. The principal consumers of offshore software work, the industrialized nations (e.g. USA, UK, and Germany), have already witnessed many manufacturing industries shift offshore as they have matured. These industrial migrations accelerated since World War II: steel, shipbuilding, automobiles, manufacturing, textiles and apparel, consumer electronics, tool making, semiconductors, and others. In the automobile industry, for example, during the period of accelerated decline of the North American industry to Japan, the market share of US firms declined dramatically from 85% in 1974 to 56% in 1991. The common denominator of these historical migrations is that, until recently, offshoring occurred in physical goods; offshoring has taken a new turn in that it is now taking place in *services*.

A useful way to understand the context of these offshoring waves is via Vernon's classic model called the *international product cycle*.[2] The model has three stages. In Stage I, a new product begins with highly skilled entrepreneurial activities, typically in industrialized nations. In Stage II, production begins to shift offshore via investments in low-wage nations. In Stage III, as the product standardizes, it is mass produced with cheap, low-skilled labor. The model seems to describe software offshoring fairly well, helping to explain the recent accelerated pace of software offshoring. Interestingly, while some software segments have now entered the third and "final" stage of the international product cycle, other software segments are still in Stage I or II.

Offshoring is still a small portion of the global software and IT services marketplace – comprising at most 5% of expenditures. A UN report dramatically labeled the new offshoring trend as a "new international division of labor" and emphasized that it is still at its early stages.[3] Where will it go? How far will it go?

Possible future trajectories are plotted in the graph of Figure 1.2 using flat "S curves." One trajectory is becoming visible: a split between design activities (high-level) and

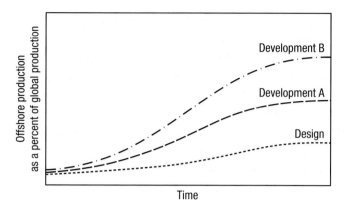

Figure 1.2 Scenarios of offshore migration.

development activities (low-level) that are migrating offshore at different rates. Offshoring will reach some plateau (though we cannot say when). The plateau may be at a lower relative level (as in "Development A") or at a higher relative level (as in "Development B"). While we cannot predict the future, we can draw six lessons from previous offshoring waves that we detail below.

1　To reiterate, in previous offshoring waves, production shifted to lower-wage nations. Some of these waves were gradual increases over time, some had an inflection point where offshoring accelerated, others moved through some kind of "S curve" reaching a new plateau (e.g. automobiles), as in the two upper lines of Figure 1.2. In consumer electronics, production – and much of the design – moved first to Japan and then shifted to the newly industrialized countries (NICs: South Korea, Taiwan, Singapore, and Hong Kong), and then to other Asian countries (China, Thailand, and Malaysia) with even lower wages. This migration pattern is referred to by international economists as the *Flying Geese Formation*, where the lead goose is the US, followed by Japan, and so on. These flying geese are beginning to appear in the software industry, as well. After the US lead, the first geese were the three Is: India, Israel, and Ireland. The second tier of geese appeared when Indian firms began to move some software work to another tier of low-wage nations: Vietnam and China.

2　In previous offshoring waves, it was not just production that moved offshore but the *know-how* about that production, as was the case in the steel industry or, separately, the semiconductor-manufacturing industry. In software offshoring, production know-how transfer is most evident in the quality standards such as Capability Maturity Model (CMM) and International Standards Organization (ISO) (which are introduced later in this chapter). The Indian organizations, and later software firms in other nations, embraced these standards and are now global leaders in their application.

3 Offshoring is a significant industry tremor leading to massive restructuring, namely: acquisitions, consolidation, job displacement, and the emergence of global giants with a broad presence in major markets. Since 2000 this tremor and its after-shocks has been evident in the IT services industry where the distinctions between the major IT services organizations in the US, Europe, and India have begun to blur. American firms in this segment increasingly resemble Indian firms in their offshore offerings, while the large Indian firms are vying for the largest contracts just like the American firms.

4 In previous offshoring waves there was often a corresponding rise in the industry's productivity in the home countries, due to a rise in R&D investments, automation, and production efficiencies. In parallel there were significant changes in the design approaches used in each industry. Charles Simonyi,[4] one of Microsoft's first software architects, argues that offshoring is but a prelude to software automation and mechanization. There is evidence that this is already taking place as software service companies scramble to automate labor-intensive tasks in data centers, software customization, translation, web site hosting, and reuse of code.

5 In some industrial offshoring waves there was a split between higher-level design activities and lower-level production activities, as in the distinction between *design* and *development* of Figure 1.2. Indeed, in the case of software, one of the forces of offshoring is standardization, allowing some factory-like approaches in software production. This is a departure from the practice of many decades in which software was practiced largely as a craft. Standardization is less evident in higher-level (design) tasks, which are more creative tasks, and which are usually the sources of a company's competitive advantage.

6 The political dynamics surrounding previous offshoring waves suggests that protectionist policies, such as import barriers, can help to slow offshore migration for some periods, but do not seem to be effective in the long term. This is an interesting lesson for industrialized nations struggling to deal with the ramifications of offshoring (see Chapter 12, Offshore Politics). The political dimensions have also changed in this offshoring wave. In the case of the US, the political constituencies of business and labor have diverged. Large corporations were vocal when the competitive threats came from Japan in the 1980s. However, in the software offshoring debate of post-2000, US firms continue to dominate the global marketplace. Not only do they not lobby for protection, to the contrary, they lobby against protectionism. The other political constituency is software labor, which in the US is largely non-unionized.

We offer a final observation in our look into the future. Offshoring will likely accelerate the formation of two industry configurations: networks and supply chains (see Figure 1.3). On the one hand, offshoring has created truly global networks of software activities, similar to the well-known network structure of the Internet. A network is set of connected nodes with each node connecting to many other nodes. It is not unusual anymore

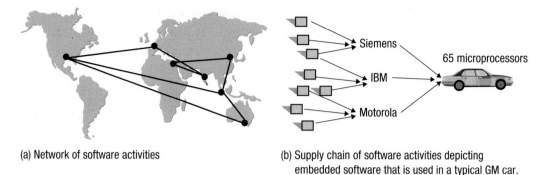

(a) Network of software activities

(b) Supply chain of software activities depicting embedded software that is used in a typical GM car.

Figure 1.3 Future structures of global software activities: network and supply chain.

to find a network of collaborating teams, as in the case of an EDS project that had such network collaboration between Mexico, Australia, Egypt, and Brazil.

On the other hand, borrowing from another business area, the software industry is beginning to resemble the auto industry in that there is a "global supply chain" of software producers, where each producer adds value as the software is transformed and then passed from one phase to the next. We see this illustrated in the auto industry itself, in the embedded software that goes into today's cars. A typical GM vehicle has about 65 specially built microprocessors, each with its own embedded software (in fact, together, these 65 microprocessors are now more expensive than the costs of all of the other raw materials that go into the car). GM writes little of this software in-house, with the exception of the microprocessor for the power train. Instead, it contracts with three major suppliers, Siemens, IBM, and Motorola, who in turn, source from a network of American, European, and Asian software centers. In short, a global supply chain of software that goes into your car.

The Offshore Stage Model: progression and diffusion

We now turn to look at companies that are offshoring in order to understand the progression and diffusion of this phenomenon. The Offshore Stage Model, first described in an article by Carmel and Agarwal,[5] helps us to tell this story.

Companies tend to move through four offshoring stages depicted in Figure 1.4. Companies that do not offshore are in Stage 1, "Offshore Bystander," in which they metaphorically watch the others. In fact, as we later discuss, most companies, whether large or small, are still in Stage 1.

Stage 2, "Experimental," is a transition stage in which companies test the offshoring waters for a year or more. For large corporations this stage's expenditures could be as large as 10–20 million USD per year. Experimentation is a wise approach for organizational learning and risk reduction because of the many difficulties in offshoring. Savvy

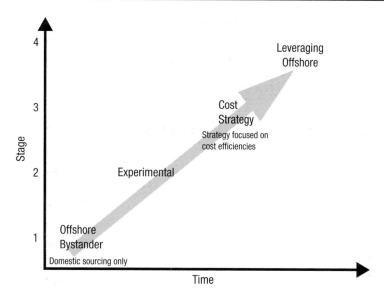

Figure 1.4 Offshore Stage Model.

managers experiment to the point where they see measurable, positive results, and only then do they grow to the next stage. Some call this the "Start-Small" strategy and, according to one study, 63% of companies are using this approach.[6]

In Stage 3, "Cost Strategy," companies begin to experience significant and consistent cost savings in their IT work. By this stage, firms have corrected some early missteps and have expanded their offshore activity as measured by number of projects, staff, or budget. There have been hundreds of firms, if not thousands, large and small, which claim cost savings in their software related activities driven by the low wages in offshore nations. Various studies have tried to determine just how much offshoring saves. The composite of studies indicate that the cost savings ranges from 15% to 40%[7] for companies offshoring at least a year (this is discussed in Chapter 2, Offshore Economics and Offshore Risks).

Experienced companies move to Stage 4, the highest stage, where they truly leverage offshoring. In this stage companies move beyond mere cost savings derived from wage differentials and benefit from other strategic advantages. Here, offshoring is used to drive innovation, speed, flexibility, and new revenues.

The Offshore Stage Model is also useful to measure offshoring *diffusion*. Since it was introduced in 2002, it has been used to estimate the ratio of large companies at each stage of the offshore progression, as shown in Table 1.1. The rough estimates in this table, made by two American research companies, indicate that only 10% of the largest US corporations were active in offshoring in 2003–2004 (i.e. they were in either Stage 3 or 4). Furthermore, about half of the largest American firms do not offshore at all. In spite of the enormous attention to offshoring in the US in the early 2000s, offshoring was still rather limited.

Table 1.1 Offshore stages of US Fortune 1000 firms

Stages	Percent of 1000 largest US firms in this stage (2003–2004)		Percent of all software work which is offshored for a typical firm in this stage (%)
	Meta Group (%)	Forrester (%)	
Stage 1	55	50–60	0
Stage 2	33	25–30	5
Stage 3	8	5–10	10–30
Stage 4	4	<5	40–50

Source: Estimates by Meta Group[8] and Forrester[9]

The stages can also be used to anticipate offshore diffusion for large firms. If, assuming conservatively, only 20 firms a year move out of Stage 1 and into Stage 2, and the annual advance from stage to stage is just 10% of the firms in that category, then by 2010, nearly one-third of US "Fortune 1000" firms will be active offshore users in Stages 3 and 4.

Strategic advantages

IT offshoring has been driven primarily by executives' desire to lower operational cost. This is the Cost Strategy of the Offshore Stage Model. Lowering operational costs does not necessarily translate into a company's strategic advantage, just as saving money on a new office lease is not a strategic advantage, but merely the relentless day-to-day effort of any company to reduce its operating costs.

However, in some industries, IT offshoring is beginning to be viewed as a *strategic necessity*. Some call it "offshore or die." When one company's cost efficiencies allow it to lower prices or expand its competitive options, then other companies must match their competitor's strategy, or fail. Offshoring is becoming part of the larger context of *hyper-competition*: companies are swept into faster and faster cycles of competitive responses and reactions in order to remain financially viable and cost competitive. *Not* offshoring may well become a strategic peril. Such was the case of one of America's largest television manufacturers, Zenith Electronics, which resisted offshoring for decades, while slowly shrinking, before it disappeared completely.

While cost reduction is the primary strategic focus of most companies that are offshoring, it is not the only strategic advantage to offshoring. The fourth and final stage in the Offshore Stage Model is labeled "Leveraging Offshore." As we saw in the estimates of Table 1.1, there are relatively few companies that have reached this stage. Those that have progressed to this stage have moved beyond mere cost reduction and benefit from innovation, speed, flexibility, and new revenues. We discuss these benefits in greater detail in Chapter 5, Offshore Strategy. Here we introduce the two most important of these additional strategic goals: attaining speed and accessing talented labor.

The first strategic lever is the increase in *speed, agility, and flexibility*. This means that companies that offshore can rapidly ramp-up (by reducing the time to get the project started) and reduce project duration (time-to-completion). The abundant supply of labor offshore gives companies greater agility: to assign a large number of engineers to a problem; to forge ahead in several directions instead of just one; to ramp-up (scale-up) and respond to a business need within days instead of months.

Companies that develop software products benefit from the second strategic advantage: *accessing talent.* For these companies, their success stems from innovation and their innovation capabilities come from their talent – their most brilliant and creative engineers. Firms that expand abroad to tap this talent are called "knowledge seekers"[10] and tend to behave somewhat differently than those seeking mainly lower wage rates. In previous decades technology companies would tap foreign talent by going to other high-wage, industrialized nations. In the 1990s, they began turning to Israel, India, and later to China. For example, by 2003, 77 global software product firms established direct R&D subsidiaries in India.[11] Many others perform contract R&D on an out-tasking basis in India.

Follow-the-sun

Stories about offshoring often mention *follow-the-sun*, also known as *round-the-clock*. Along with low costs, follow-the-sun is another allure of offshoring. It is often mentioned by those who seek to make offshoring sound unique. Follow-the-sun, as the name hints at, exploits time zone differences to speed up project work. For example, a team in America can hand off its work at the end of its day to team members in India or China, who can then continue the work while the US team members sleep.

This has undeniable appeal. If software work can be coordinated properly, then project duration can be reduced by a factor of two. Moreover, if three teams are correctly positioned across time zones, then a theoretical threefold duration reduction is possible. This is much like a factory running three shifts, 24 hours per day, producing three times the volume. Using follow-the-sun development, a company may be able to save months from the development cycle and release a product earlier, thus giving it a competitive advantage. This is an enormous potential benefit of offshoring.

However, coordination in follow-the-sun must be flawless in order to reduce project duration. One miscommunication can delay the entire day's worth of work. In practice, few globally dispersed software efforts have been able to fully capitalize on the theoretical advantages of follow-the-sun. Daily follow-the-sun coordination is simply too difficult for software teams. An IBM team, described in Carmel's 1999 book,[12] was set up to capitalize on follow-the-sun. However, fairly quickly, the global team discovered that daily handoffs were too difficult to coordinate.

Nevertheless, follow-the-sun can be effective for some activities and for certain phases in software work. Startups in Silicon Valley have been excited about rapid prototyping of new software products in which the coding is done in India, and then sent back to the US for comments and refinement. Activities, such as bug-fixing (in the maintenance phase), or call-centers (e.g. technical support), are better suited to follow-the-sun, because they are usually small tasks (low granularity), of low complexity, and can be routinized between the time-separated sites.

Offshore challenges

Recall that roughly half of America's largest corporations are not offshoring at all. The percentage is higher in Europe, and higher still once small- and medium-sized firms are included.

Why is this? Why is it that relatively so few American, European, and Japanese firms are offshoring? All this is in a business environment in which offshoring is one of the accepted, if not expected, strategies. For example, the US strategic consulting firm BCG issued a report in 2004 that practically shouted, "the real question now is not whether to go global, but how much and how fast you can move."

There are many reasons for the relatively small participation in offshoring, but we begin with a simple one: it is more difficult to work with people far away than those close by. It is more difficult, because of five factors introduced here (and covered in more detail in Chapter 8, Overcoming Distance and Time):

1 *Communication breakdown.* We human beings communicate best when we are close. Yet, offshoring is all about working with people far away, with whom communication is conducted via "narrow" channels such as e-mail or telephone. A software engineer would always want to conduct a difficult design session face-to-face. Why? Because people communicate with more than mere text or words. The way the words are delivered (via tone of voice, the pauses in speech, the body language, the gesturing at the whiteboard) are all vital. Some say that 80% of the messages we convey go beyond the plain text. The all-too-frequent result of communication over distance is that dreaded word: miscommunication.

2 *Coordination breakdown.* Software is a complex task that requires many small and large adjustments. People who work on a common task coordinate via countless adjustments: a question, a request for clarification, a small improvement, an *ad hoc* solution resulting from a 1-minute chat while standing in line at the cafe. So much coordination comes from spontaneous, face-to-face conversation. When offshoring, all of these small adjustments do not take place, certainly not easily. When coordination slows or breaks down, several dynamics occur. Problem solving gets delayed again and again, or the project goes down the wrong track until it becomes very expensive to fix.

3 *Control breakdown.* Successful management control takes place when managers can roam around to see, observe, and dialogue with their staff. Hence, management by walking around (MBWA). When a team leader or project manager

is supervising software developers many kilometers away, roaming around and getting a "feel" for what's happening becomes an unusual event. Sometimes it never happens at all. And, when managers cannot roam, they have to rely on collecting information and imposing their will by means of technology: telephones and e-mail. This is less effective than face-to-face.

4 *Cohesion barriers*. Groups that are close together jell and bond. People get to like each other, trust each other, help each other, and work harder for each other. Offshoring introduces a situation in which the group of dispersed individuals is unlikely to form these tight social bonds.

5 *Culture clash*. Offshoring means going to far-away lands and working with foreign cultures. Each culture has different principles, values, beliefs, communication norms, and behaviors that are embedded deep in our minds. In fact, we now understand that our respective cultures are "programmed" into our minds by age 10. The result of all these deeply engrained differences is that in any cross-cultural communication, the receiver is more likely to misinterpret messages or cues. Hence, the familiar complaint of *miscommunication across cultures*. We devote Chapter 9 to the problems and solutions of cross-cultural communications.

These five factors represent only *some* of the difficulties that make offshoring difficult. These and other offshore challenges translate into the extra costs of offshoring. These extra costs can sometimes offset wage advantages making offshoring a losing proposition. The extra costs of offshoring are covered in Chapter 2.

What is done offshore?

> *"... not a single activity is immune ... [to offshore]."*
>
> Findings from a 2003 survey
> by the American industry magazine
> "Software Development"[13]

> *"Offshore [workforce] is less innovative ... the technological innovation stays here [in the US]."*
>
> Director of offshore development,
> US embedded software company

> *"Everything can be done offshore except for when you're developing new hardware and the software for it – at the same time."*
>
> Director of an offshore development center,
> Motorola

There is no consensus on the question of whether all software activities can now be done offshore. There is consensus, however, that there are certain activities that are a better fit at offshore locations while others are better to leave in-house and in-country – "onshore."

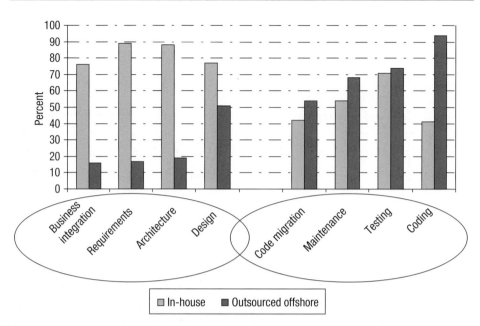

Figure 1.5 Offshoring by phase of the software life cycle.
Source: Software Development Magazine, 2004.[14]
Survey results of US development managers/engineers in projects that were already partially or
fully offshored. Numbers do not add to 100 because the activity could be both in-house and
offshore and because some projects did not use some phases.

The best method for analyzing which activities are suitable for offshoring is to use the
software development life cycle. This was the approach taken in a survey conducted by an
American trade magazine that is depicted in Figure 1.5. The survey results show that
activities offshored most often were coding, testing, and maintenance; while those that
were least offshored were business integration, architecture, and requirements gathering.

This is a picture of two clusters: activities that are frequently offshored and those
that are not. This clustering is essentially the *split* offshore migration that was described
earlier in the chapter (Figure 1.2), in the forecasted separation between design and development. Recall that the chart forecasted that development (such as coding) would move
offshore faster than design (where the word *design* was used broadly to include any
high-level activity including architecture). The survey results are consistent with the styl-
ized depiction of the life cycle in Figure 1.6, applied to offshore and onshore.

Let us take a closer look at these two clusters to see what differentiates them from
one another. First, activities offshored are those which: are standardized (commoditized),
can be precisely defined (precisely specified), and may even be considered tedious, repet-
itive, and undesirable. Hence, coding and maintenance, which can be well defined, are
often offshored.

Activities that tend to stay onshore are those which require customer interaction,
customer proximity, deep domain knowledge, and deep cultural knowledge. Thus,

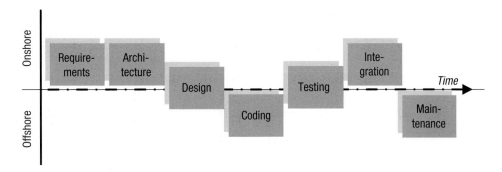

Figure 1.6 Common division of onshore/offshore phases.

requirements gathering activities, early in the development cycle, are conducted onshore because of the need to be close to customers; that is, to meet with them and talk to them in their own language. At the end of the development cycle comes systems integration (also labeled "deployment"), where the software pieces are sewn together; this is difficult to do from afar.

Two phases appear as both onshore and offshore in Figure 1.6: design and testing. High-level design is often done onshore while lower-level design is offshored. Testing, a largely tedious task, is offshored in part. Unit testing can easily be done offshore. Some testing activities, such as systems testing, have remained onshore, because they are better performed together with other integration/deployment tasks. Separately, for control reasons, managers keep later testing phases onshore in order to monitor quality.

Some important IT activities are not depicted in Figure 1.6: data center management and a variety of managed services, such as network services and support. Generally, these have not been offshored since they require some proximity. Furthermore, there are business continuity risks in moving such services offshore.

The final differentiation between the two clusters is *competitiveness*. Companies tend to keep "high-end" activities at home (onshore) to maintain their competitiveness. These tasks are more likely to be creative, innovative, and research oriented, or they require very broad knowledge and experience. For example, architecture (systems architecture or product architecture) fits all of these criteria and is kept onshore as shown in Figure 1.6.

Companies give these competitive activities many labels namely, core activities, proprietary activities, the most sensitive activities, the leading product design, and the company's competitive edge. Most IT applications at end-user firms (e.g. a bank) involve few activities which are proprietary and therefore such companies are not hesitant to offshore. The picture is different, at least in theory, for a software product company: its software code is its "crown jewel." Any company should carefully weigh allowing an outsider access to its crown jewels. Yet even among software product firms that perform R&D there is significant offshoring. An estimated 5–15% of American software product R&D is offshored, forecast to rise by 2007/2008 to 25–30%.[15] And some of the companies which are offshoring software product R&D are the largest US technology firms. Yet, this is a risk that they have presumably considered carefully. We return to discuss this topic in Chapter 5, Offshore Strategy.

Quality, processes, and methodologies

Until the 1960s, Japan was known as a destination for cheap, low-quality manufactured goods. "Made in Japan" was a derogatory term. Beginning in the 1950s, the Japanese began to implement quality production processes from abroad and re-invented their production culture. Within a relatively short period, Japan became known as a source for high-quality manufacturing.

This is an example of history repeating itself. Until about 1990 India's reputation was for shoddy products. The Indian offshore providers recognized early that by attaining high-quality development processes (CMM, ISO 9001, and Six Sigma) they could achieve two important goals: they could overcome some of the difficulties of working over distance, and, perhaps more important for the nascent Indian industry, they could signal to potential customers that they were world-class companies worthy of their business.

The Indian organizations' success in this regard is remarkable. The very first organization in the world to attain the highest-level software process maturity rating, the CMM Level 5, was India's Motorola unit in 1993. Since then many other Indian firms have adapted this standard. By 2001, 32 organizations with CMM Level 5 were in India. By 2003, there were over 65 Indian organizations that attained CMM Level 5 (including both Indian firms and foreign multinational subsidiaries), representing more than half of all organizations in the world at that highest level. The outcomes of all these efforts have been impressive: a 2003 survey among US offshore users found that 71% of the users stated "that offshore suppliers delivered somewhat better- or much-better-quality work" compared to their US-based counterparts.[16]

By the early 2000s, offshore units in nations that compete with India, namely China and Russia, began to emulate the Indian sector's successes in this area. Russia's Luxoft attained CMM Level 5 in 2003, while two Chinese organizations attained Level 5 by 2003.

CMM and its newer cousin CMM-I, both came out of the US Software Engineering Institute to impose structured, standardized development processes that are repeatable, planned, and optimized. The advantages of these process improvements are error rate reductions and cost savings. Moreover when working across distance, these processes reduce the communication difficulties by structuring the development process – by introducing standard approaches that reduce the need for explicit coordination.

Nevertheless, several issues need to be noted about the promise of structured approaches in general, and CMM in particular. CMM cannot address many of the offshore challenges that were described earlier in this chapter. Furthermore, the fixation on CMM Level 5 is misguided: CMM Level 3 is considered sufficient for most software development activities (CMM Level 5 was designed to satisfy large American defense contractors and NASA). Finally, the hoopla around CMM generated by the Indian success has created an environment of exaggeration and fraud. A 2004 CIO magazine[17] exposé documented this landscape: some offshore providers misrepresent their CMM rating and others have not "renewed" their ratings in years since there is no requirement for such renewals.

National differences in software abilities

When in a comfortable private setting, executive decision-makers ask the following questions: Are the programmers in Country A better than the programmers in Country B? Do the engineers of Country C have a better work ethics than those of Country D? Are the firms in Country E more sophisticated than those in Country F?

These are important questions. No one has answered them properly. And it is unlikely that anyone will, in part because of the extreme political sensitivity of giving a country a failing grade. Furthermore, comparative studies are likely to be fraught with sample biases. Project outcomes are often mixed and depend on many factors including shortcomings of the client. We found an American software manager that set out to answer the "which country is better for us" question on his own, using an interesting approach. His small software firm set out to contract an identical pilot project to three providers, one each in India, China, and Russia. Once the pilot projects were complete the firm chose the winner. The case appears in Chapter 4.

An important study was conducted by Cusumano and colleagues[18] using data collected in 2001–2002. The study compared best practices of firms in Europe, Japan, USA, and India (most were technology firms). The eleven best practices included: architecture specifications, code generation, code reviews, sub-cycles, pair programming, and daily builds. The researchers concluded that India's stated strengths in processes and best practices are correct, namely, that Indian firms were combining disciplined approaches such as architectural specification, code and design reviews, with more flexible approaches. The study also noted continued software strengths in Japan. Separately, the study authors made an interesting observation in their conclusions, writing that "… no Indian or Japanese company has yet to make any real global mark in widely-recognized software innovation, which has long been the provenance of US and a few European software firms."

The demand for offshore work

In the next two sections we examine the two sides of the offshoring equation: the *demand* (answering the question: who are the customers?) and then later the *supply* (answering the question: who are the providers?). The global context of demand and supply is presented in Table 1.2.

On the demand side we begin with a few broad observations before looking at specific geographic regions:

- *Geography.* The most aggressive offshore consumers are in the US. Within Europe, the UK has been the most active in offshoring.
- *Industry.* Among the end-user industries, the most active in offshoring are financial services (banks, investment firms, and insurance) and technology firms (software, hardware, and telecommunications).[21]
- *Company size.* Companies that offshore are generally the larger firms, with the largest global corporations, such as GE, American Express, and British Telecom having taken the lead. The vast majority of small- and medium-sized firms (SMEs) do not offshore, with the exception of technology firms.
- *Motivation.* Cost savings has always been the dominant driver for offshoring in both Europe and America, except for the short period in the late 1990s, when labor

Table 1.2 Selected figures and estimates on market sizes

	Data for 2003	Forecast for 2008
Global market of IT services (*Source*: Gartner)	536 billion USD growing at 6.2%	N/A
Global market of IT-enabled services (*Source*: IDC)	405 billion USD, growing at 8%	680 billion USD
Indian IT services providers (*Source*: Gartner)	Growth of 29% for the year, though Indian providers represent only 1.4% of global market	N/A
Foreign R&D subsidiaries	India: 77 global software product firms.[19] China: 223 technology multinationals.[20]	N/A
Indian R&D sourcing (*Source*: Frost & Sullivan)	1.3 billion USD	9.1 billion USD
Global [offshore] sourcing of software and services (*Source*: ITAA)	10 billion USD. Total savings from offshoring by US corporations was 6.7 billion USD	21 billion USD. Total savings from offshoring by US corporations will be 31 billion USD

markets were very tight due to the technology boom, the Y2K remediation efforts, and the euro currency conversion.

USA

The largest market for offshoring is the US. This should not be surprising given that the US has long maintained leadership in the software industry through its clusters of innovation, such as Silicon Valley, its catalyst role on the internet, and its powerful global corporations. The largest of these corporations, IBM, may qualify as the "Father" of globalized software development. In the 1960s and 1970s, IBM, then the most powerful computer firm in the world, and always among the top 10 largest US corporations, had R&D centers in three countries with development centers in some additional European nations (see Figure 1.7). No company came close to this global network for many decades.

In 2003, global sourcing of computer software and services was estimated at 10 billion USD, which represents but a small portion of total US corporate spending on IT. Of the 10 billion USD in global sourcing, the US purchases have generally represented two-thirds. This ratio will probably remain steady for some years, though the amount in dollars will continue to rise at double-digit rates.

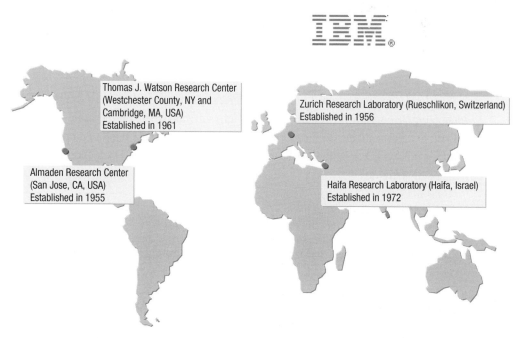

Figure 1.7 IBM's early global network of R&D sites, circa 1970.
Source: IBM.

Table 1.3 Selected US software firms' offshore activities

Cadence	Cadence is the largest semiconductor design software firm. The firm opened an India R&D center in 1987 where its staff reached 300 developers by 2004. It also had development centers in Taiwan, China, and Russia[22]
Google	In 2004, search engine company Google was viewed as one of the most innovative American tech firms and its public offering of stock was closely watched. Google stated that it could afford to hire the best developers in the US, but chose to open an R&D center in India to attract creative talent
I2	I2 develops logistics software for business customers. In 2004 the firm had 1400 engineers outside the US, mostly in India, representing about half of the firm's global employees. The Indian staff included about 180 staffers of Indian origin who were previously working in the US and volunteered to go back to India[23]
Microsoft	Microsoft has traditionally been cautious about dispersion of its R&D, with nearly all activities at its Redmond headquarters. Its four foreign centers, India, China, Israel, and England, were set up in 1990s. Growth since 2000 has been primarily in India and China. The firm had roughly 300 employees in its Hyderabad (India) locations scheduled to double within 2 years
Oracle	Oracle began offshoring in 1990. By 2003, its Indian centers grew to 3000 developers with a stated goal of doubling to 6000. Oracle also performs offshore R&D in Ireland and China

The offshore spending is disproportionately concentrated in larger firms, particularly in financial services. But even among the largest firms, only a minority is actively offshoring. Estimates regarding offshoring from the largest US firms, the Fortune 1000, summarized in Table 1.1, indicate that in 2004 only about 10% of these US firms were past the experimental stage of offshoring in which more than 10% of their budgets were devoted to offshoring.

The landscape is quite different for American technology firms. All the top 20 US technology firms perform at least some offshore software work. Furthermore, at all sizes (small, medium, and large) high-tech firms that are offshoring are much more prevalent than for non-tech firms. Table 1.3 presents a sampling of some US technology firms' offshoring activities.

Europe

Europe has been slower than the US in offshoring IT, even though the cost advantages in doing so are just as pronounced. Offshoring from the wealthy industrialized nations of Western Europe is estimated at 2.5 billion USD, representing only about 25% of US offshoring volume. In 2003, the US was responsible for almost 70% of the export of Indian IT services, and Europe for only 22%.[24]

There are several reasons for the US–European differences. The European firms have a more conservative style of doing business than the Americans, taking fewer risks in sourcing and are more particular about spending on specific projects. There is a more inward style of functioning in Europe and organizations prefer to deal with vendors whom they know well. Trust is more important, and therefore building personal relationships takes a longer time. In addition, Europe has stricter labor laws with greater restrictions on redundancies (layoffs). Bringing staff from offshore firms to Europe is difficult due to visa restrictions. Finally, it is clear that language plays a role. There are few Indians who can read French or Danish manuals, fewer still who can build organizational applications with French or Danish interfaces.

Britain is the exception and is the European leader in offshoring. British firms are the preferred partners for English-speaking offshore countries, and especially for its former colonies (e.g. India, Pakistan, Bangladesh, and Sri Lanka). Ninety-five percent of all UK offshore work is from India, estimated at 1.2 billion USD in 2003. Although the UK IT services market share in Western Europe is 21%,[25] it is responsible for 59% of the Indian software exports to Europe. Indian software giant Tata Consultancy Services (TCS) set up operations in the UK way back in 1975, a time when the word offshoring had yet to be coined. It's first client was CMIG, an insurance company. By 2004, TCS had 3700 professionals working for British clients, of which 1600 were working onsite.

Indian IT is quite visible in Britain with hundreds of offices of Indian providers who are increasingly partnering with local and foreign players. Wipro works together with Accenture, at the Thames Water project; and TCS with Fujitsu, at the National Health Service. The majority of the offshore work is being performed by end-user organizations, such as in the financial services industry, telecom, and retail. By the early 2000s, large British organizations such as ICI, Lloyd's, British Telecom, London Underground, British Airways, P&O Nedlloyd, UBS Warburg, Marks & Spencer, Tesco, and Safeway were offshoring. Similar to the picture in the US, offshoring still represents only a small sliver of the huge market in IT services: 2.4% in 2002, forecast to double by 2006.[26]

And then we come to Germany. Roughly 80% of Germany's largest companies have not yet offshored.[27] German firms have been reluctant to offshore for several reasons. The first reason may well be language and culture: in order to successfully work with clients, good German language skills are essential since many German employees are unable to discuss detailed business issues in English. Second aspect is the historical background: unlike former colonial powers, namely Britain and The Netherlands, Germany has fewer ties with distant cultures. As a result, decision-makers are hesitant about foreign services sourcing. Third, Germany's *mittelstand*, its medium-sized firms that are the foundation of the economy, has been slow to offshore because of their more conservative, go-slow strategies.

Important German software companies, such as SAP and Software AG, are actively offshoring. SAP is the most successful European software product company. Its development center in India is the largest development hub for SAP outside Germany. The firm employed 1000 staff in Bangalore, in 2004, with plans to triple that number within 2 years.[28] SAP also has a smaller development center in Shanghai. By 2004, Siemens, the electronics and engineering giant, aimed to shift around 10,000 jobs from Germany to

Figure 1.8 (a) The Shanghai Multimedia Park, opened in China in 2002, houses more than one hundred multimedia and software companies. It is located in the western part of Shanghai, and is surrounded by institutions of higher learning.

Figure 1.8 (b) Part of the main campus of Infosys, one of the largest IT services firms in India. The high-tech campus in Bangalore includes development and training centers, meeting rooms, and restaurants.

lower-cost countries in Eastern Europe and Asia. The banking and finance sector is a key vertical in the German offshore market; Deutsche Bank and Dresdner Bank are large consumers of offshoring services. Other offshore clients are car producer BMW and the national airline Lufthansa. It is also worth noting that, while the UK's primary off-shoring destination has been India, Germany looks to Eastern Europe. German firms offshore about 60% of their work to Eastern Europe, with only 40% to India.[29] A 2004 poll found that the German experiences with these "nearshore" countries were some-what better than with India.[30]

Figure 1.8 (c) Headquarters of Politec in Brazil's capital Brasilia. Politec is one of the largest IT services firms in Brazil.

Figure 1.8 (d) BaliCamp is an offshore development center on the tropical island of Bali. It is designed in traditional architectural style on a mountainside. BaliCamp was built in 1998 by Sigma, a major Indonesian IT services company.

Smaller European countries, by virtue of necessity, have a more international outlook, are multi-lingual, and are more capable of communicating in English. An example is Switzerland, where offshoring corporations include Swissair, Nestlé, and banking multi-nationals Crédit Suisse and Union Bank of Switzerland. In Scandinavia, offshoring corporations include Swedish telecommunications equipment maker Ericsson and mobile phone producer Nokia from Finland.

The Dutch have been offshoring for more than 20 years. Jan Baan, the technology entrepreneur, illustrates the long Dutch history of offshoring. While still a small software firm in the late 1980s, Baan established an Indian development center. As he built his software company in the 1990s into a global giant he expanded the company's Indian centers significantly, reaching a peak of 1000 employees in India, by 2000. Once Baan, the company, collapsed, Jan Baan started Dutch software firm Cordys in 2001, which quickly set up offshore operations with a 300-staff development center in Hyderabad, India.

Over the years, at least 250 Dutch companies have offshored, including Philips, Shell, ABN Amro Bank, and KLM. As with other nations, it is the large Dutch multinationals that moved into offshoring the fastest. But some small firms have looked offshore as well. For example, Decos is a small Dutch software company with 20 employees, producing applications for document management. In 2000, it started a subsidiary in Pune, India, with 12 staff, and now sells offshore services.

The Dutch have also been egalitarian in choosing their suppliers: we counted at least 35 nations that have conducted software work for Dutch organizations, including its former colony of Surinam, and some unlikely spots such as North Korea and Iran. India is the preferred source, and the volume of Indian software exports to Holland is around 100 million USD. In 2004, we estimate that 5000 offshore staff were working for various Dutch projects. This is a modest amount, since a total of 250,000 people are employed in Dutch IT functions. A 2004 study estimated that this number could grow tenfold in 10 years to 50,000,[31] which would represent a significant volume for such a small country.

Other countries

The offshore opportunities have rippled to other nations outside the traditional industrialized economies. Two cases in the Middle East are noteworthy. Israel is both a destination for innovative offshore R&D and a nation that sources some software tasks to India. The largest Israeli software firm, Amdocs, in 2003 announced the opening of an offshore development center in India. Other large Israeli firms soon followed suit. The wealthy Gulf nations have long relied on foreign (mostly American) contractors to build and maintain much of their national IT base. With globalization, many of these activities have been picked up by regional firms in India and in Pakistan.

The offshore supply

Close to 100 nations are now exporting software services and software products. The buyers are now presented with an "offshore menu" to satisfy any taste. The new offshore destinations span the economic categories: from newly industrialized economies, through transition economies, to developing economies and even some least-developed nations. Much of the volume of the recent wave of software offshoring is going to three large nations, the "Big Three" nations: China, India, and Russia. The common denominator of these three nations is that they have very large populations (which is why we label them as "Big"), and each of these nations is endowed with a large, well-educated work force in science and technology. The "Big Three" offshore nations' software exports are displayed in Figure 1.9, along with the exports from Israel, an important offshore software destination, but certainly not large in its labor pool.

In the "Big Three" offshore nations, India, China, and Russia, there were over 1000 organizations exporting software products or services in 2004. Add to this number hundreds of firms in destinations from Brazil, through Romania, and east to the Philippines.

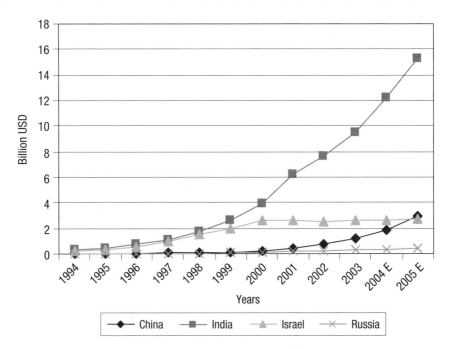

Figure 1.9 Export of software products and services from major offshore destinations.

These numbers do not include the micro-firms of three or five programmers, which are too numerous to tabulate. By comparison, The Netherlands has a few hundred software firms and the US has 2200 software product firms[32] and roughly 4000 IT services firms.[33]

As we noted at the beginning of this chapter, the offshore software labor pool has grown markedly. In the "Big Three" offshore nations, India, China, and Russia, the 2004 workforce may be as high as half-a-million inside those organizations that are exporting software and IT services. This number does not include those engaged in projects at home, which is significant in China. The labor pipeline is also large in these three nations, producing tens of thousands of engineers and computer scientists every year. Add to the "Big Three" tens of thousands of software professionals in other offshore nations, and the result is a very large offshore labor pool, seemingly endless. In the US, for comparison, the number of software employees in 2004 was 675,000 according to the US government's official statistics[34] (though quite a bit larger according to broader definitions).

The result of the expanding global supply is fierce competition driving down prices. This is good for the buyers in the industrialized nations. Two indicators illustrate price competition: one for larger offshore deals and the other for very small deals. First, at the upper end is the deal structure, where only a few years ago, according to IDC,[35] 85% of contracts were billed on an hourly basis, with resulting higher costs for the buyers. By 2004, spurred by competition, this ratio fell dramatically, with only 20% of contracts billed on an hourly basis, and with many of the remainder priced using some performance incentive.

Then we come to the low end of the software marketplace. For very small projects online programming marketplaces match IT buyers and sellers in reverse auctions. A business that puts up a small project for bids often receives 5–15 bids from around the world. The fierce price competition resulting from offshoring is palpable: American programmers are competing with programmers from India and other offshore nations, in bidding on small programming jobs for as little as 10 or 25 USD per project.

The rise of Indian providers and the competitive response

One of the reasons that offshoring expanded so rapidly was the ability of the Indian IT industry to grow large firms quickly. These are the IT services firms, commonly called *providers*, which are the suppliers of contracted services in outsourcing (or out-tasking) engagements. Since 2000 the top Indian providers, now labeled "Tier-1" firms, and numbering 5–7, have transformed themselves from Indian firms into globally competing, full service companies. In other words, these Indian-based providers have become technology *multinationals* that are less India-centric.

The growth and geographic expansion of the top Indian providers has been remarkable. TCS, the largest of the Indian-based providers, exceeded 1 billion USD in revenues by 2003, and had quickly expanded its delivery centers around the world as illustrated in Figure 1.10 (delivery center is the label for development center). Since 2000 TCS and

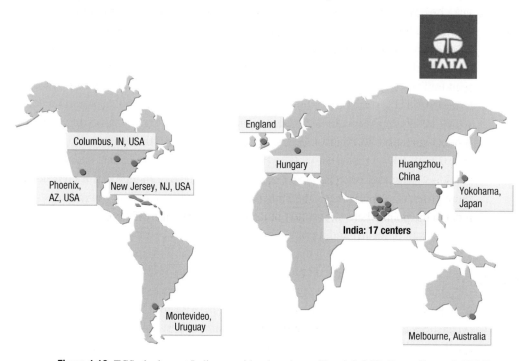

Figure 1.10 TCS, the largest Indian provider; locations of its global "Delivery Centers" (2003).

other Indian Tier-1 providers began moving into Eastern Europe and China to head off competition and establish bridgeheads to a range of customers. Some of this expansion was achieved by hiring more staff in the US and Europe, while other expansion was conducted through acquisitions, such as the Indian provider Wipro taking over the energy services division of US-based AMS in 2003.

Observing all this with great concern are the non-Indian providers, particularly the largest of these firms (in North America: IBM, EDS, CSC, and HP; in Europe: Capgemini, Xansa, LogicaCMG, and Atos Origin; and in Japan: Fujitsu). Indian Tier-1 providers are threatening to these global giants just as small, nimble, and low-cost airlines are threatening the traditional, stodgy airlines in North America and Europe. The Indian providers began winning deals and growing at a breakneck speed of 20–40% per year, while the growth rate for the non-Indian providers was a moderate 2–7% per year.

The competitive response of American and European giants has been to compete head on with the Indian providers by rapidly growing their own offshore centers and becoming, in effect, offshore providers. Anglo-Dutch firm LogicaCMG started a Bangalore (India) center in 1998 growing it to 1200 staff by 2004; declaring that it will double in size within a year after that. The firm also has two smaller offshore centers in two other low-wage countries (Czech Republic and Malaysia). Atos Origin, a French-based provider, built an offshore network including centers in India (its largest offshore center), Poland, Hungary, China, Malaysia, Brazil, and Argentina. The American giants expanded even more. For example, EDS had 13 offshore centers in low-wage nations in early 2003, growing to 18 by the end of the year. Giant IBM has 16 offshore centers.

Remarkably, these large Western providers look more and more like the top Indian firms: they are presenting their clients with hefty offshore menus, in effect inviting them to choose any offshore location, as if selecting a vacation destination at a travel agency. The difference between the large providers, be they American, European, or Indian, is blurring. In terms of sheer size, for example, the top three Indian providers, all at 25,000–30,000 employees, have exceeded most of the European providers, though still far from IBM's 175,000 employees in its IT services division, or even the some-what smaller Accenture at 95,000 employees. The blurring of Indian and non-Indian providers is not limited to the global giants: medium-sized firms have also become off-shore providers. For example, the French company Valtech expanded aggressively into India to the point where almost half of its 800 employees are offshore.

Nearshoring

In some ways nearshoring is the *opposite* of offshoring to India. Many companies prefer destinations that are close by, but still less expensive that at home. This is why, in offshore-speak, Canada, with its lower-wages, has become a nearshore destination for US firms. The proximity in time zones, travel time, and culture makes Canada easier to deal with.

Except for Britain, many European nations seek suppliers closer to home: the Finns look, at least in part, to Estonia and the other Baltic states; much of the Germans' offshoring is nearshore to Poland, the Czech Republic and Hungary, and other Eastern European nations; the Italians sometimes nearshore to Yugoslavia.

Japanese companies nearshore to China, with some activity in Vietnam and Korea (though due to relatively high costs in South Korea, it has fallen out of favor). The city of Dalian in northeastern China, with its historic ties to Japan, has a relatively high concentration of Japanese offshoring. NEC has been among the most active in China. The firm began to develop software in China in 1982 and outsources software work to 40 Chinese firms which employ over 3000 programmers. Several other large Japanese firms were reported to be spending 10–30 million USD annually in China.

Some consumer nations, such as the US and the UK, do not nearshore much of their work. For example, the US is the principal software customer for India, Russia, and Israel; all are countries far away. Similarly, the British generally prefer India. However, some offshoring from the US does stay nearshore in the American continent: Canada was mentioned earlier; the Mexican industry has been marketing itself as the ideal nearshoring destination for US firms; and other work is spread all over Latin America.

IT-enabled services

American lender E-loan gave customers two choices when they called in for loan application processing. If they want the loan processed the same day, they should press 1 for the Indian center; if they want their loan processed in the US, which may take longer, they should press 2. The company reported that 86% of the customers pressed 1.[36]

Apart from software-related work, a very large number of other activities can be performed offshore: customer interaction services, such as telemarketing, helpdesks, and call centers. They also include various types of back office work, such as market research, tax preparation, airline and hotel reservations, insurance claims processing, financial research and Human Resource functions. They also include professional services in the field of data and content integration, such as medical transcriptions, data entry, digitizing, animation and multimedia. Finally, they include engineering services, such as computer aided design (CAD) or architectural drawings.

All these services have one thing in common: they are substantially dependent on IT. And since an IT infrastructure is an important tool and enabler to perform this work offshore, the generic term IT-enabled services (ITES) is used to describe this diverse range of activities. The abbreviation BPO (Business Process Outsourcing) is also used, but this mainly refers to the back office activities. IT-enabled services offshoring is not a new phenomenon: credit card processing in India for American customers has been taking place for two decades, though it has accelerated only since the early 2000s.

An example of a large IT-enabled services user is HSBC, the UK-based financial services company. In 2003, it had 1500 employees in China and 2000 in India to provide clerical processing and call-center services. More than half of those employees in China and India service UK accounts. The firm also opened a new 500-seat processing center in Kuala Lumpur, Malaysia.[37] In 2004, it announced plans to transfer thousands of jobs in the UK to India and Malaysia. It is also looking at Sri Lanka and the Philippines to mitigate country risks.

Growth in IT-enabled services is spurred by falling telecommunications costs, particularly in contact centers (often referred to as call centers). Today, it costs almost nothing to have a client use a toll-free number to call India or the Philippines. Contact centers, where wages account for the majority of the costs, are now being moved offshore on a massive scale. US-based Delta Air Lines launched a plan in 2003 to cut operating expenses by 15% by the end of 2005 and moved parts of its call-center reservations operations to India. This move will save the company more than 12 million USD.[38] British Telecom opened a large call center in India in 1999. French speaking call centers are set up in North Africa and Mauritius. A Dutch-speaking helpdesk operates in South Africa.

Many of the information processing services are labor-intensive and involve small and repeatable tasks, such as processing a telephone call, or a single tax form, or a drawing. These routine tasks can be peeled away, since they do not require proximity, and then done offshore, cheaply. Since it is relatively easy to standardize and structure this work, communication with the offshore site is less complicated and results in fewer extra costs. Relative to IT, the cost savings for IT-enabled services are much easier to benchmark and can be higher: IT-enabled services cost savings of 40% and even up to 80% are reported.[39] GE, one of the pioneers of outsourcing service operations to India, reported annual savings of 340 million USD per year from its Indian operations.

Offshoring IT-enabled services is closely tied to offshoring IT. IT-enabled services and IT are similar in that they both are driven by declining connectivity costs and both are part of the new burst of international trade in services. But the ties are stronger still. Many of the offshore IT providers are also IT-enabled service providers. Many of the most aggressive corporations offshoring IT in the US and Western Europe are also aggressive in offshoring IT-enabled services. Finally, many of the countries offering IT services have also grown IT-enabled services. Major Indian providers such as HCL and TCS have built IT-enabled services operations. Major global IT providers have grown their IT-enabled services: IBM bought Daksh e-services, India's third largest back-office services firm, in 2004.

A large number of countries are active in IT-enabled services offshoring, and the choice of locations may be even larger than with IT. Indian is the leader in this area: according to NASSCOM, there were 210,000 people working in exporting IT-enabled services in India in 2004, with revenues of 3.6 billion USD. Extraordinary growth forecasts call for IT-enabled service revenues in India to reach 64 billion USD by 2012. Multinationals like Bank of America, British Airways, Swissair, American Express, ABN Amro Bank, GE, and Citibank have all set up captive facilities in India. GE alone has 12,000 people working in its many Indian IT-enabled service centers.

The Philippines IT-enabled services exports, at 600 million USD, are now twice as high as its IT exports. Filipino IT-enabled services centers are used by ChevronTexaco, American Express, Procter & Gamble, and Accenture. Central European nations that entered the EU in 2004 are preferred destinations for many continental Europeans. For example, Poland has IT-enabled services centers for Fiat, Lufthansa, and Philips. Even African nations tend to be more active in IT-enabled services. These include: South Africa, Ghana, and the Indian Ocean nation of Mauritius. Interestingly, security cameras in some US car-parks are monitored remotely from the small West African nation of Cape Verde.[40]

In some cases the risks are higher when offshoring IT-enabled services relative to offshoring IT. When offshoring application development work some projects get delayed because of unclear specifications or miscommunication. This can be irritating or even problematic, but it is usually not critical to the company: business carries on. This is different with IT-enabled services. When outsourcing revenue-generating processes offshore, such as contact centers, there is risk to an entire business process. Revenue can be lost if customers are frustrated with the offshore service provider. A widely published example is Dell, which decided in 2003 to redirect some customer service calls to helpdesks in the US, rather than to its call center in Bangalore, India. Dell brought some activities back to the US when it found that several of its business customers complained that Indian technical support workers relied too heavily on scripted answers and could not handle complex computer problems.

IT-enabled services has already become "the next big thing" in offshoring. The American consultancy Forrester predicted that by 2015, 3.3 million American service jobs will move offshore, most of these in IT-enabled services.[41] We return to the topic of IT-enabled services when we examine the efforts of nations to build their offshore IT industry in Chapter 10.

Concluding comments

Now that we have introduced the rich landscape of offshoring, let us take you back to an observation we made earlier in this chapter. Recall that we introduced two evolving configurations of global software activities. The first is that of a global *network* in which nodes of software producers are interconnected. The second is a *supply chain* structure where each software producer adds value to the software and passes it on the next phase in the production process.

These offshore configurations bring us to the title of this book. As offshoring diffuses, the software industry will increasingly resemble other industries in which components and services are sourced globally from high-, medium-, or low-wage nations. Over time there will be less that is unique in "offshoring." Instead, what we label today as "offshoring" will become simply "global sourcing." Perhaps our next book may have this title.

2 Offshore economics and offshore risks

Three providers bid on the state of New Mexico unemployment system. Two American firms, IBM and TRW bid 12 million and 18 million USD, respectively. The winner, Tata Consultancy Services based in India, came in at less than 6 million USD.[1]

Solidcare Systems, a Security Software Startup in Silicon Valley, has no research and development (R&D) in Silicon Valley. Instead, to stretch its 5.3 million USD in venture capital funding, all of its software engineers are in India earning much lower wages.[2]

Philips, the Dutch electronics giant, faces strong competition from Asian producers. It claims to be saving millions of dollars by moving major parts of its technical and embedded software development to cheaper locations. Its Indian software center now houses 900 employees and is responsible for 20% of Philips' worldwide software content.[3]

There are hundreds of such firms, large and small, all of which trumpet their offshore costs savings. Companies in the wealthy industrialized nations are dazzled by offshore programmers' low wages. Since cost savings are the dominant reason for offshoring, much of this chapter is devoted to dissecting the controversy regarding the actual costs and savings. Then, later in the chapter, we turn to the myriad of risks which companies face when offshoring.

Labor arbitrage: finding the lowest wages

A software development manager is shopping for the lowest-cost labor suppliers and is drawn to the low-wage offshore nations. Thus, he is acting somewhat like an arbitrage financier: he is sourcing labor where it is the cheapest to use it, where it can earn the greatest return. Since we are dealing with software, in which there are few expensive assets like factories, the costs of development are driven mostly by the wages of software labor, from the junior programmers to the seasoned project managers. For example, variations on Table 2.1, showing comparative wage levels of software professionals,

Table 2.1 Wages for software professionals. Annual, in USD

USA	63,000(d)
Australia	62,000(a)
Canada	57,000(a)
UK	45,000–99,000(e); 81,000(a)
Japan	44,000(d)
Singapore	43,000(a)
Israel	39,500(c)
Ireland	23,000–34,000(f); 35,000; 23,000(h)
Brazil	20,000(a)
South Africa	18,000(f); 30,000(a)
Mexico	7000(h)
Philippines	5000(h); 6600(d); 10,000(b)
India	5000(h); 5900(d); 9000(a)
Russia	5000(h); 5500–7500(f); 7500(d); 7000–18,000
Indonesia	5000
Ukraine	5000(b)
Poland	4800–8000(f); 9000
Pakistan	3600–6100(f)
China	3000(h); 4700(d); 7000–14,200
Romania	2300(f)
Vietnam	1400–6000; 3000(h)

Sources: Multiple.[5]

have appeared frequently in the IT business press.[4] The wage differentials are striking when viewed by managers being pressed to reduce costs.

The software engineers in these offshore destinations are not being paid undignified rates. To the contrary, the Indian programmer's basket of goods and services that she can purchase with an annual income of 9000 USD is roughly equivalent to her European counterpart. The cost of middle-class housing is much less in Bangalore than in Silicon Valley. These equivalences represent the notion of purchasing power parity (PPP), a traditional economic index, reflecting different national costs after currency conversion.

Wage data of Table 2.1 need to be treated carefully since they vary considerably – by as much as 50%. Wages are substantially higher in the principal cities: higher in Shanghai than in Suzhou, higher in Moscow than in Novosibirsk, higher in Bangalore than in Ahmadabad, higher in London than in Wales, and higher in Silicon Valley than in Omaha. Wages also vary considerably depending on the size of the firm, with engineers at large firms typically earning 25% or more than engineers at small firms.

A number of sources try to be consistent across countries by focusing on wage data for junior programmers, typically fresh out of university, since this is the most objective comparison. However, the salary differences (compared to programmers in the US and Europe) tend to be most pronounced for junior programmers, and the salary gap gets compressed for more experienced engineers and managers. For example, wage differential between Europe and India is roughly 1:8 (Europe:India) at the junior programmer level and only roughly 1:3 (Europe:India) at the high managerial levels.[6] Thus, many wage comparisons in the press bias the wage differentials of offshore labor.

Wages are only part of the story. Companies that are managing their own offshore centers need to be looking at the "loaded" wages that professionals receive. Direct wages reflect only part of the labor costs. Benefits are thick and elaborate in many nations. In India benefits include mini-bus pick-up and drop-off every work day. In Russia, where consumer finance is undeveloped, firms give loans to buy apartments. In Israel they include a company car in more than one-third of the cases, as well as some lunch discounts in about two-thirds of the cases.[7] In China benefits include an extra 13th month payment for the lunar holiday, a housing allowance at 17% of wages, medical insurance at 12% of wages, and more. A leaked IBM internal memo shows that a Chinese programmer's cost, including salary and benefits, is 12.50 USD an hour, still only about 20% of an equivalent American IBM programmer (the comparison was for a programmer with 3–5 years experience).[8]

Once other costs are factored, then the *fully burdened* Indian employee costs about 30,000–40,000 USD per annum. This includes the cost of rent, which is relatively high in the major Indian destinations (1 USD per square foot in good locations), support, services, insurance, taxes, and other items. For comparison, loaded cost (including benefits and taxes) for an American software engineer, at a large company, is about 120,000 USD and a burdened cost is about 150,000 USD per annum.

We have heard from more than one naïve Western manager that Indian wages will soon be bid up to reach those in the US and Europe. This is misinformed. With its large pipeline of fresh engineering graduates and its enormous underclass serving as a drag on wages, Indian software engineering wages will not approach US or British wages in our lifetime unless there is a catastrophic economic depression in the West that does not touch Asia (an unlikely delinking). That said, wages in India and China have been bid up for quite a few years, rising 10–30% per year. One result of the strong demand is that India is no longer the lowest cost software nation. Instead, some firms (including some Indian firms) are turning to China, Vietnam, and others, where wages are lower.

Clients that are not directly issuing payroll checks to their offshore employees are paying a *charge rate* to the offshore provider. The charge rate is much higher than the wage rate and tends to hover around 15–35 USD per hour for India and Russia. This rate is higher than the wage rate because it includes the provider's overhead, marketing, training, and profit (these profits can be high: the Indian firm Infosys is the most profitable major software services company in the world). The charge rate for larger Indian firms is at least twice the wage rates and usually several times than that. However, these rates tend to fluctuate quite a bit, based on general supply-demand conditions and based on the customer's negotiating power. Charge rates below 10 USD per hour have been unusual for the larger (Tier-1) Indian firms, but are more common for the small firms. American providers using offshore resources do their best to compete with the Indian providers, but generally offer somewhat higher charge rates than their Indian competitors. Nevertheless, we have heard of very low charge rates offered by the American providers using offshore resources: for example, an HP bid for market share in 2004 at a charge rate of 15 USD per hour; a large US firm bid for 10 USD per hour for maintenance work.

Then we come to the *onshore charge rate*. At least some of the labor needs to be performed onshore to be close to the client. This labor is charged at a different, higher rate, ranging from 35 to 80 USD per hour. At the upper range this onshore rate is close to a fully loaded rate for a US-based engineer. The onshore professionals need to be charged at higher levels, since they incur higher costs and sometimes may need to be paid the prevailing rate for legal reasons.

For some years the offshore providers were emphasizing the "80:20" ratio in their marketing pitches: only 20% of the staff need to be onshore charging high rates, while 80% of the staff are offshore charging the more attractive lower rates. In practice, this ratio varies considerably by phase, project, and client. We know of many clients that maintain the ratio at about 50:50. Why would clients choose a financially disadvantageous ratio? Some clients are sluggish in transitioning work offshore (as discussed in change management in Chapter 7) since reaching higher ratios offshore requires proactive policies. Other clients simply viewed their offshore (primarily Indian) providers as suppliers of labor, only at lower costs. In its extreme form this is *body-shopping*, the label that is shunned by all involved in the offshoring business because it is suggestive of exploitation.

Labor arbitrage is most dramatic at the bottom end of the global programming landscape. Companies from countries such as Bangladesh, Nepal, or Vietnam are asking very low rates, sometimes below 5 USD per hour. Another source of very cheap rates is the assortment of online programming marketplaces that pit programmers against each other as they compete for small projects. These online marketplaces[9] work using a reverse auction mechanism in which the lowest bidder is more likely to be selected by the client. In a study of these marketplaces conducted by Carmel and Espinosa, a key finding was that in head-to-head competition between providers from industrialized nations (e.g. the US) and offshore nations (e.g. India), the median winning project bid amount was only 35 USD for the whole project.[10] This is a strong indication that offshoring is driving down prices, since programmers from high-income nations cannot survive on proceeds from such tiny projects.

Transactions Costs and Total Savings from Offshore Strategy

In spite of the low offshore wages, total offshore costs can actually be higher than before offshoring. Why? Because of the economic concept of *Transactions Costs*.[11] This concept helps to explain the economics of offshoring. Classical economics was based on the theory of efficient markets and thus it should always be cheaper to contract out than to hire. Yet, by the 1900s it became evident that this theory was deficient, since it could not explain why giant corporations appeared to be quite successful at creating vertically integrated companies that seemed to do nearly everything within the boundaries of the firm.

Thus, the Theory of Transactions Costs encompasses a powerful idea: that there are real costs incurred when going out to the marketplace for identifying suppliers, negotiating and contracting with these suppliers, and then, later, policing these suppliers so that they produce to the desired quality level.

Now we apply this to offshoring. Companies that are acting in their own self-interest look to the marketplace *if* their production cost savings of outsourced offshore work outweigh the additional Transactions Costs incurred when dealing outside the firm. In the case of offshoring, most of the production cost savings are due to the lower wage costs for the software staff in low-wage nations. Thus, to put it numerically, it is more efficient to offshore when:

Production cost savings $>$ Sum of all Transactions Costs

Of course, the difficulty when offshoring is that the Transactions Costs are difficult to assess without experience or benchmarking. Furthermore, offshore Transactions Costs tend to be higher when the tasks are difficult to define, when uncertainty is high, and when complexity is high.

In order to capture the notion of Transactions Costs, the outsourcing and offshoring industry coined two useful terms, namely *TCO*, the Total Cost of Offshoring; and *TCE*, the Total Cost of Engagement. Both of these terms attempt to capture all the costs of offshore activity in order to compare them with "onshore" or "normal" costs. The problem with TCO and TCE is that many firms end up offshoring using the "Stumble-and-then-Succeed" form of offshoring. Here firms go offshore and encounter problems and failures. They stick with their offshore decision and eventually make it work.

Such was the case of Silicon-Valley-based ValiCert which later became part of information security firm Tumbleweed.[12] Following the contraction of the technology crash of 2000, ValiCert struggled as a company. In 2001 it hired Indian-based Infosys, laying-off programmers in the US in this process. After numerous difficulties including its own lack of experience, the company created its own Indian subsidiary reaching 60 employees. During this phase ValiCert continued to experience problems and strains between the Indian and US offices. Only in 2003, more than 18 months after offshoring began, did this small global company begin to properly perform in a distributed structure. At that point the company attributed its very survival to the Indian subsidiary's cost savings. It had computed that the burdened cost of its Indian employees was roughly one-sixth of the cost of its Silicon Valley-based employees. Thus, ValiCert practiced "Stumble-and-then-Succeed."

In sum, given the extra Transactions Costs and the many cases of Stumble-and-then-Succeed, the correct measure for offshoring should be Total Savings of Offshore Strategy (TSOS).

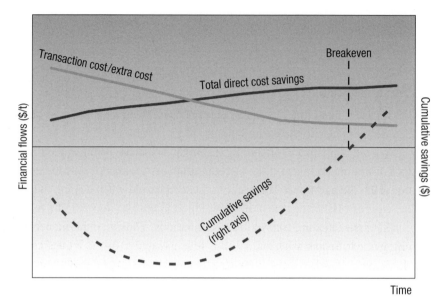

Figure 2.1 Offshore benefits, costs, and savings over time.

By using the notion of TSOS, a portfolio of projects is examined over a certain time period. Some of these, on an individual basis, will show negative savings, as would any project portfolio. For new users, the first projects are always hard to manage and cost savings will only come later. Successful firms will show a positive TSOS over time. Put differently, this is the "learning curve" that companies need to climb up. The word *strategy* is used in TSOS because there must be a multi-period offshoring strategy in order to achieve economic benefits.

Figure 2.1 depicts a typical cost and benefit stream over time, with cumulative savings rising up in a classic inverted-J-shape, also known as a "hockey-stick." The financial milestone is the breakeven point where the cumulative savings begin to be positive. This is the TSOS. Of course, the firm's time period to achieve a positive TSOS will vary. Small, well-specified projects (out-tasking) can reach positive TSOS quickly, sometimes immediately. More significant engagements are unlikely to reach positive TSOS in less than 1 year – and may often be 2 years and more. And, of course, in some cases a positive TSOS is never achieved.

Extra offshore costs

Transactions Costs is the more formal term for the grouping of costs which we call *extra offshore costs*, and which has also been labeled *hidden costs*.[13] While labor arbitrage makes offshoring attractive, these extra cost components tend to muddy the picture. The extra cost items are listed in Exhibit 2.1. Each of these cost items is discussed in detail in this section.

- Search and Contract
- Restructuring
- Infrastructure: technology and connectivity
- Knowledge transfer
- Efficiency
- Travel
- Overhead allocation
- Governance
- Mitigating risks (e.g. disaster recovery)

Exhibit 2.1   Extra offshore cost items.

The first category, *Search and Contract*, is usually a one-time cost for the first engagement. Search and Contract costs for outsourcing revolve around provider selection. These costs include initial research, consulting fees, legal, contracting, and travel. For some companies, this item is small because their search is opportunistic; for example, they contract with a provider they met coincidentally. The duration of this phase is usually months, and sometimes approaches a year. The overseas trip by one or more executives has become part of the ritual of offshoring and is a part of the search costs.

Another initial cost is *restructuring*. This is the euphemism used to signify layoffs, severance pay and any retention costs: paying the essential people extra so they will not leave. Retention expenses are critical because firms need to keep the key employees to share their knowledge, facilitating the knowledge transfer (KT) described further below. This cost item is more common on the American side of the ocean than in Europe, where offshore projects start small and layoffs are less common.

The rest of the cost items are mostly ongoing costs, although they change over time. *Infrastructure costs (technology and connectivity)* are typically small and fairly predictable. The offshore unit may need to procure hardware and software. In some cases clients have to purchase dual equipment and software licenses: one set for onshore and the other for offshore. There are additional difficulties: software licensing issues may be convoluted and more expensive offshore; and in India and in other developing countries, long lead times are required to procure some equipment.

The connectivity picture is better, costs are falling rapidly. International call rates that were once significant are no longer burdensome, and voice over IP is beginning to eliminate voice costs altogether. But voice lines and multi-party conference lines are still needed in many cases. The annual leased costs of E-1 (2 Mbps) and DS-3 (45 Mbps) circuits from India to US/UK are 43,000 and 780,000 USD, respectively (2004). Thus, the cost of basic connectivity, an E-1 circuit, is roughly equivalent to the charge for just one offshore programmer per year.

The most difficult category to forecast is the cost of Knowledge transfer (KT), sometimes called *technology transfer*. KT is the notion of moving specific knowledge and experiences into the *minds* of the offshore developers. While some of the KT involves well-understood skills and rules, much of it is in *tacit knowledge* – knowing what is in-between the lines of a software specification. Packing up the software specifications and "throwing them over the wall" works for simple cases only. It does not work for most cases. IT managers recognize this, although in practice they often fail to manage the KT process properly. A more extensive discussion of KT is found in Chapter 7.

The largest KT cost may be the redundancy that is built into the project early on. A typical scenario takes place when several offshore developers need to travel to the client site for KT early in the project life cycle (which happened in the T-Corp case described later in this chapter). At these early stages the client firm is paying for double staffing for the same work, for its current and the offshore employees. KT shows up in various other ways: for non-English-speaking countries, translation costs need to be added. Another way to look at KT is to anticipate that things will go wrong: poor KT early in the process leads to costly rework and repair. In the case of outsourcing offshore, some KT costs may be absorbed by the provider, such as with KT problems that stem from high turnover rates, cultural training for provider employees, and some training related to transitioning.

Efficiency is a productivity ratio comparing the onshore, "original" unit's efficiency (the baseline) to that of the offshore unit. During the first few months the new offshore individuals are less productive as they "go up the learning curve," but over time their efficiency rises as they master the knowledge and skills required. From a cost perspective, however, software efficiency is difficult to measure, let alone forecast. While it is possible to measure efficiency using objective measures such as Function Point per dollar, this is rarely practiced. In contrast, measurement is relatively easy for offshore call centers using measures such as number of calls per hour, duration of calls, and call satisfaction.

Some examples illustrate how the efficiency item can be used in practice. Firstly, how efficient is the new offshore unit at the beginning of the project? Sand Hill Group estimates 24% offshore efficiency within 2 months; while US-based T-Corp (described in the case later in this chapter) used a 50% efficiency for an offshore maintenance team after 4 months of KT before any production work actually began. Restating these numbers, this means the offshore personnel are only 24% or 50% as efficient as the baseline (the original software personnel) within 2 or 4 months, respectively, from the project start date. These low efficiency ratios early in the offshore cycle are simply due to the normal need to learn the complex knowledge.

Can the offshore units reach 100% efficiency and when? Again, we use these two examples. Sand Hill Group posits that efficiency increases rapidly in the first several months, but never quite reaches 100%. Similarly, T-Corp does not forecast reaching 100% of the baseline efficiency. T-Corp assumes, based on experience, that the offshore personnel will reach 95% of the baseline efficiency by the end of the first year.

Others claim that offshore efficiency actually rises above the 100% baseline due to qualified people and solid processes offshore. This is the assumption used by Gartner,[14] which uses the term "effectiveness factor." This factor is a composite of several components, namely technological expertise of the offshore unit, its project management expertise, and its business-domain expertise. Gartner calculates that this effectiveness factor is lower for the typical US "Fortune 1000" corporation than for the typical large American and Indian providers. In other words, the providers are more effective than the client. In fact, Gartner estimates that once a steady state is reached, they are 50% more effective: a significant difference!

Due to KT and other needs, *travel* costs are significant. Yet, they tend to be under-budgeted. For example, if there is a need for extended onsite work, an offshore team of six developers visiting the client site onshore for 3 months will cost about 150,000 USD.[15] On top of the direct cost is an obscured organizational cost associated with travel to far-flung locations – the opportunity costs of many wasted days in airplanes, jet lag recovery, and sick days due to exotic food and water. Most offshore projects expect travel to take place at the beginning or the end of the project life cycle, or sometimes both. Of course, travel costs vary. For example, they are lower if the work is done nearshore.

The last three cost categories are *Overhead, Governance*, and *Risk Mitigation*. Overhead allocation, an accounting item, varies from company to company and may sway the economic benefits of offshoring from positive to negative. Governance costs represent about 5–10% of an offshoring contract. Governance costs include new positions to communicate with the provider and to carefully monitor the provider's work by collecting and analyzing data. Another buried cost item includes contract management costs such as handling invoices and payments for outsourcing. Finally, risk mitigation is the investment in resources in case of failure, such as backup and recovery.

The last remaining issue is the overall impact. What is the total of all of these extra offshore costs? Not surprisingly, the rough estimates (in Table 2.2) indicate a very high variance of these extra offshore costs: from 12% to 57% of the contract amount.

Table 2.2 Extra offshore costs (%)

	Meta Group[16]	CIO magazine (composite)[17]
Search and Contract[18]	1–2	0.02–2
Restructuring (layoffs and retention)		3–5
Communication	1–3	
Process changes (KT)	0–10	1–10
Cultural differences (KT)	2–5	
Transitioning the work (KT)		2–3
Lost productivity/cultural issues (KT)		3–27
Efficiency	0–20	
Travel	2–3	
Governance	5–7	6–10
Turnover at offshore site	1–2	
Total extra costs	**12–52**	**15–57**

Each item represents additional cost (%) of the overall offshore outsourcing contract. The four items marked knowledge transfer (KT) include KT as at least part of that item. "Lost productivity" includes turnover at offshore unit.

In other words, if the extra costs are kept under control and managed closely, they will be smaller than the wage savings, and lead to overall offshore savings. However, at the upper range of these estimates, at 57%, the labor savings are wiped out and offshoring ends up costing more.

What is the bottom line? Does offshoring lead to cost savings?

So far, we dissected the main economic offshoring trade-offs: the wage differentials versus the extra costs of offshoring. This begs the important question: Are the extra offshore costs indeed smaller than the wage differentials? After all, so many firms have reported offshore savings.

Four consulting firms estimated this bottom line and concluded that offshore savings are positive. Firstly, a study by US-based Deloitte Consulting[19] finds that in the best case of offshore outsourcing savings are in the 25% range when considering all costs and benefits. This savings level can be achieved by a typical "Fortune 50" US corporation that has been offshoring for 5 years. Deloitte calls this "as good as it gets." More typical success cases are Fortune 50 companies offshoring for at least 2 years and saving 15%. The study author speculates that if an experienced firm optimally combined all best practices and processes that it would return a theoretical 47% total savings. A second study, by US-based Sand Hill Group, estimates savings for software product R&D firms at 40% of R&D budget,[20] adding that Return-On Investment is reached in a year. Third, Gartner estimates the typical offshore outsourcing savings for large firms to be in the 28–40% range. Fourth, US-based Magnolia Communications surveyed New York City companies that offshore and reported that their savings were 44%, although the study authors noted that savings of nearly this total are possible by simply moving to a less-expensive city in the state of New York, such as Syracuse.[21]

These estimates and surveys should be viewed with great caution because their methodologies are not rigorous. We have not seen studies that have examined a broad range of companies' offshore strategies and produced a comprehensive comparison of cost savings. We are skeptical that such a study can reasonably be done because of the difficulty of standardizing assumptions and overhead rates.

In any case, the various studies have a limited bearing for the case of any specific company, because cost savings are not guaranteed by statistics. Nevertheless, the thrust of this chapter should have made clear that the *extra offshore costs* are not hidden costs at all. They are only *hidden* costs for those companies at early stages of offshoring with little idea of what to expect. They are *known* costs which can be identified, decomposed, and most importantly, managed. Managing the process is the key. If the process is well managed the TSOS will likely be positive. One 2004 study conducted in Europe found that 80% of companies "suffered problems" in offshoring.[22] This is hardly shocking since "normal" software projects suffer problems 64% of the time according to the well-known Standish surveys of project success.[23]

An important lesson is that cost savings are heavily dependent on time. In other words, many of the extra offshore costs decrease with time. This is the positive impact of organizational learning illustrated in the "Stumble-and-then-Succeed" story of ValiCert earlier in this chapter. There were many wrong turns, frustrations, and wasted spending.

Yet another lesson of the cost savings computations is that if your firm is offshoring it must first produce good internal benchmarks to determine if you are indeed saving money. And you need good cost accounting to compute these benchmarks. As you continue to expand offshore you need to show *real* cost savings in order to move forward.

As a final note, all of these financial analyses ignore two vital issues: strategy and risk. Firstly, the analysis, thus far, covered only costs and cost savings, yet the benefits of offshoring can be in less quantitative benefits: in strategic and tactical advantages; in speed; in quick ramp-up time; in availability of able resources; and innovation. The strategic benefits of offshoring are discussed in Chapter 5. Secondly, even a comprehensive cost analysis does not address risk. In other words, the cost savings projections could show substantial savings, but the risk factors may be too high. Cost and risk are not the same. Offshore risks are covered later in this chapter.

Case study Calculating the extra offshore costs at a giant American corporation

The case of T-Corp illustrates the process of offshore cost computation in detail. The case is an actual case, but at the request of this large American company, all identifying details are disguised.

The Finance Officer took a copy of the offshore spreadsheet and computed the financial net present value (NPV) for the proposed offshore engagement. It exceeded the 15% that the division sets as a minimum threshold for budgetary approval. "This is great," remarked the Finance Officer, cheerily, and blessed the project.

Bobby Sanders directs the Global Services unit at T-Corp, a US "Fortune 500" technology company. Global Services is an internal division tasked with matching the corporation's internal units to offshore resources. Bobby manages a network of six captive offshore centers (wholly owned by T-Corp); in India, Singapore, and several other locations. Much of the software code developed at T-Corp is *embedded* software.

In 1998, Bobby developed an offshore spreadsheet template to assess the financial benefits of offshore work and assist his internal corporate customers in making offshore decisions. Between 1999 and 2004 he used the offshore template to assess 55 candidate engagements. Bobby noted with pride that about one-third of the proposed projects were rejected by "running the numbers." In these cases the "economies simply weren't there" and the decision was made to leave these activities onshore. Some of these were rejected because they were end-of-life projects where the expensive knowledge transfer process was not justified.

One of Bobby's most important "wins" for the Global Services unit was in 2000, in persuading T-Corp's Strategic Software Division (SSD) to begin to peel off some of its lower value work offshore.

As its first foray into offshore, SSD decided to consider the work of 30 engineers at its Ohio engineering site. These were engineers that were immersed in their embedded code. They knew it, they built it, and they maintained it. But every modification request (MR) that came in tended to distract the engineers from their most important task – working on the next product release. Quite simply they were falling behind. As they fell further behind, they became more attracted to the offshore pitch: "We like this offshore idea," Bobby remembers SSD's Director saying.

Donald Robert, one of SSD's product directors, visited Bobby's office to begin evaluating the offshore engagement. The two sat down in front of the computer screen and Bobby pulled up his offshore spreadsheet template and began to explain how it can be used to help in making the offshore decision.

The first hurdle in using the offshore template was lack of benchmark data. The Ohio product group had no process data that could be used as a basis for computation. The group collected almost no metrics. Given this, how could they make the financial case for offshoring? Through dialogue Bobby and Donald found a reasonable proxy. They examined the engineers' time sheets and then estimated the amount of time the engineers spent on MRs. This estimate came to 33% of their total work time. Bobby and Donald used this figure to compute the baseline figure in the spreadsheet. Thus, 30 engineers multiplied by the fully burdened cost of 150,000 USD per year $\times$ 33%. This resulting number, 1.5 million USD, became the current *on*shore cost. This figure was entered as the first computation item into Part 1 of the offshore spreadsheet representing onshore costs.

The next important issue was making an estimate about knowledge transfer, which knowledge transfer was going to be expensive and time-consuming because the current 2 million lines of embedded code had almost no documentation. This was taken into account when an important "plug factor" was used to drive the spreadsheet – which Bobby calls an *efficiency factor*. Clearly, the offshore engineer was not going to perform at 100% of an Ohio engineer's capacity from Day 1. Bobby usually uses a 50% efficiency for the first few months of the engagement. That is, each offshore programmer is only half as efficient as his onshore counterpart. But as knowledge transfer proceeds successfully, offshore efficiency goes up. In this case, Bobby used a gradually increasing efficiency factor ending the first year at 95% and continuing at 95% for subsequent years.

The other major offshore cost items are listed here:

- Onsite training involved bringing five of the Singaporean engineers to Ohio for 2 to 5 months for knowledge transfer. The costs of apartment rental, airfare, and per diem for this period was not cheap: 92,000 USD.

- Fully burdened offshore labor costs were 5400 USD per month.
- Infrastructure expenses were broken down into three categories, some of which are driven by tax considerations rather than "straightforward" economics. The first item is non-recurring infrastructure (an "expense-able" item), principally software and hardware purchases, which usually include customs and tariffs. When pointing at this, Bobby complained that software licenses at the foreign sites tend to be higher than in the US. The second item for infrastructure is mostly ongoing costs such as leased communication lines. The final infrastructure item is infrastructure subject to depreciation which is beneficial with respect to foreign tax treatments in some countries, such as Brazil.
- The last cost item was one of the largest, the local (onshore) resources. This is the redundancy that was built in at the beginning of the engagement. The Ohio engineers would have to continue working on the MRs while the new offshore engineers were learning and acquiring knowledge in the first few months; in other words, while the offshore engineers were still far less than 100% efficient. In the first few months this item, local resources, represented a substantial cost item at over $50,000 per month, dropping quickly after that as the offshore engineers become capable and more efficient.

Bobby pointed out how well knowledge transfer was managed; it was budgeted correctly in the offshore spreadsheet and, more importantly, it was managed well. Four full months of US engineers' time were budgeted, at 45,000 USD, before the engagement even began. The Ohio division paid close attention to the details of knowledge transfer, such as on-going job enhancement of the offshore engineers, helping to keep turnover at the offshore site to a manageable 10%.

In 2004, Bobby speculated that the offshore engineers, now with 4 years of experience, were actually operating at 120% efficiency. In other words, they had become more efficient than the Ohio engineers who had trained them. He explains this by pointing out that the Singapore (offshore) engineers, who do nothing but MR work, stay more focused. But, Bobby noted that he has never gone so far as to use a spreadsheet efficiency factor above 100% for the offshore units. "It's just too speculative" he said, as he shrugged his shoulders.

After some weeks of data collection and many telephone calls, the offshore spreadsheet for SSD was complete. It showed a positive cumulative savings by Month 18. Bobby's offshore spreadsheet produced cost summaries and cost savings, but no NPV. SSD's Financial Officers took a copy of the spreadsheet and computed the NPV. The NPV exceeded the 15% that the division sets as a threshold. "This is great", remarked the Finance Officer, cheerily, and blessed the project. Bobby already understood the finance game by then. He always avoided submitting projects in the third and fourth quarters because they could not show payback by the end of the fiscal year and were likely to be rejected by Finance.

The offshore effort represented a long-term investment in knowledge transfer, as can be seen in Figure 2.2, showing the actual forecast out to the end of 2004 – an almost 5 year timeline from engagement launch. The projection shows a total cost savings of 2 million USD over 5 years. Four years into this engagement, the forecast was deemed valid and SSD's management was pleased with the offshore impact on productivity. In early 2004, to validate the economics of offshoring, SSD collected and analyzed data from a large number of MRs handled in either Ohio or Singapore. The analysis revealed that at a US engineer's rates (fully burdened), it cost 7000 USD to fix an MR, but only 4500 USD to fix an MR in Singapore.

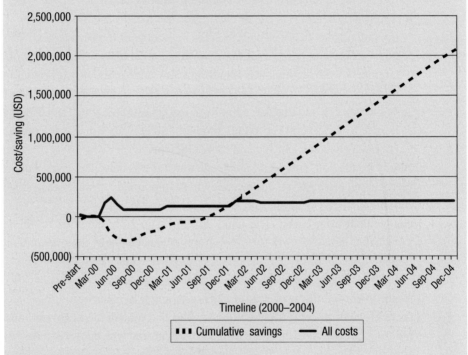

Figure 2.2 Actual consolidated spreadsheet graph for T-Corp Ohio SSD case. The large upward sloping curve represents cumulative savings.
Source: T-Corp, as developed by managers before offshore engagement.

Offshore risks

Any cross-border business increases risks; this is a constant in doing international business. Risk is the *uncertainty in doing business*. Offshore risks should not be seen in the same category as the largely predictable "extra" offshore costs, covered earlier in this chapter, which result from various factors such as difficulties in knowledge transfer, cultural communication problems, and increased coordination overhead. Offshore risks are the surprises. In business we do not like surprises. This section focuses on those risks, listed in Exhibit 2.2, which are greater because the work is done offshore.

The most important of the offshore risks is *country risk*, which is introduced here and expanded upon in Chapter 4, The Offshore Country Menu. Country risk is a broad, umbrella term, encompassing within it *political risk* and *financial risk*.
Offshoring is about working in developing or emerging countries, which have historically been more volatile, less stable, less predictable, and less transparent. Offshoring software work exposes the company to a myriad of issues in the host country.

The consequences of country risk can be severe because they affect, among other things, *business continuity* – the ability of the firm to continue its core operations. When offshoring, companies are exposed to increased risks of war, terrorism, rioting, uprising, confiscation, expropriation, and currency crises. These are real risks. They do happen. We just do not know how to predict them, especially the big, shocking events. A number of companies measure and rank these country-level risks, but they have been shown to be quite poor at predicting the big surprises – the crises, such as the Argentinean collapse of 2001–2002, or the Asian financial crisis of 1997. Therefore, these country risk assessments are of limited use in evaluating offshoring. One country-level financial risk is *currency risk*, the risk of exposure due to change in exchange rates. Absent a crisis, this risk can be mitigated by determining the currency of payment in the case of an outsourcer, or by currency hedging.

Another part of country risk is the risk of government regulatory changes. This is also called *sovereign risk.* A generation ago it was common to fear confiscation, nationalization, and expropriation. These seem less likely today. However, governments may change tax or subsidy structures that once favored offshore operations making them less attractive. Thus, favorable tax treatments in, say, the Philippines or China may change. Governments may also change the regulations that govern technology joint ventures in order to favor the local partner.

Once IT operations are in a country, most country risks are not *controllable*. The risk may be diversified, mitigated, or insured,[24] but it cannot be controlled since in theory,

- Country
- Intellectual Property (IP)
- Loss of proprietary knowledge
- Data security
- Corruption
- System security
- Contractual
- Infrastructure
- Societal and regulatory changes in your home country

Exhibit 2.2 Risk categories that are introduced or may be greater due to offshoring.

no single company has the power, for example, to stop a war between India and Pakistan. We heard an interesting rumor on this issue of controlling risk: during the Indian–Pakistani tensions of 2002, GE's CEO spoke to India's Premier and warned him that if war breaks out, GE will have to move most of its vast operations out of India. After a brief period of crisis, tensions subsided (see also the GE case in Chapter 5).

Large firms with substantial operations in India have reacted to the issue of country risk by diversifying to multiple nations (sometimes called multiple sourcing). IT managers mitigate the risk consequences by devoting more attention to their offshore contingency plans. Such plans may include mirrored systems or backup sites in another country, such as Singapore or Mauritius. Attention must be focused on whether the offshore unit has actually tested the backup plan. Equally important is the need for extensive system and maintenance documentation so that other staff can, if needed, pick up the work in case of crisis.

Risk likelihood and severity

When offshoring, a firm's goal is to reduce the firm's risk exposure, which is the product of *the probability of a bad outcome and the severity of its consequences*.

Consider the case of a hypothetical leading-edge Swedish software company, Björn, developing a state-of-the-art product. Björn's software code is its crown jewels. But, also vital is the know-how in the minds of the key engineers gained from working with the core software and with its specialized customers. What can go wrong here – what are the risks? Firstly, the software code can be copied and used in the product of Björn's chief rival Kjerstin-Tech. Secondly, key people can transfer knowledge to Kjerstin-Tech.

What is the probability of these two adverse events happening? This is a key question. In Sweden, Björn may be able to legally enforce its code ownership against Kjerstin-Tech. The mere knowledge of enforcement may give pause to Kjerstin-Tech. Björn may also be able to prevent some knowledge seepage through non-compete legal clauses in individual employment contracts. Beyond the legal limitations, it may also be able to control its "intellectual capital" through various "social controls," which are typically achieved by keeping knowledge in just one location.

On the other hand, in many offshore countries these restrictions and barriers are weaker or non-existent. Therefore, we can say that the probability of such an adverse event is higher offshore. It is more likely; perhaps, slightly more likely; perhaps, a lot more likely. For example, the performance-based contracts, which are increasingly common offshore, drive engineers to leave firms, exacerbating turnover, leading to increased likelihood that these engineers take with them specialized knowledge.

Next, we come to the *severity* of the event. Let us assume that the severity of these adverse events to Björn is high. If code or know-how falls into the hands of Kjerstin-Tech, it will take away 20% of Björn's market share. But it really does not matter if the thief is a Swede, an American, or Chinese. Once it happens, the severity – loss of 20% market share – is the same.

However, loss of key assets (code, knowledge) may have more severe competitive risks if a foreign firm, we call it Foreign-Soft, uses those assets to improve its products and build its base in various foreign countries, slowly encroaching on Björn's global competitiveness. In such a case, the severity of the adverse event may be even higher offshore.

Eight additional offshore risks

The risk illustrated in the Björn example is *IP risk* and this is the first of the additional risks presented here. Eight offshoring risks are introduced in this section.[25] This is not a compilation of *all* project and outsourcing risks,[26] of which many exist, but rather those risks which are either "new" because of offshoring or for which the probability or severity of them occurring may be higher due to offshoring.

Intellectual Property (IP) risk. In most of the target developing nations, enforcement of IP breaches is rare, while theft of software code or ideas (trade secrets) leaves the aggrieved party little practical recourse. Two recent examples illustrate these perils, but they also illustrate the emerging regime of remedies and enforcement. The first case, in China, occurred when one of the leading technology companies, Huawei, used some of Cisco's code, complete with comments and errors, in its own products. Huawei has become a direct competitor of Cisco in China and in emerging markets.[27] The case was settled by mutual agreement by the two parties in 2004.

The second case is of SolidWorks, a US software product firm which outsourced to an Indian provider, GSSL. A software engineer working on this software was fired from GSSL and copied the software code before he left. He then contacted several competitors of SolidWorks in an attempt to sell them the stolen code. He was caught by a sting operation conducted by the US FBI and the Indian Central Bureau of Intelligence.[28]

Loss of proprietary knowledge risk. This is a long-term strategic risk faced by some companies, mostly in software products and embedded software. Knowledge leakage far from home may be more likely or have more serious consequences than such knowledge leakage at home. The hypothetical Swedish case described above illustrates this risk. Critical know-how may trickle to competitors, who may be in a better position to capitalize on this knowledge competitively. The consequences of this risk do not appear until several years later, and thus, many managers, with their short-term orientation, may be less mindful of its consequences. This risk is also interesting vis-à-vis the public policy debate taking place in the US about its national competitiveness.

Next, several criminal risks are likelier when offshoring, or have greater severity of adverse outcomes, or both:

- *Data security risk.* Increased attention to privacy concerns regarding personal, individual data began first with the European Union (stemming from its 1998 Directive) and more recently in the US. The IT community will likely need to devote greater attention to this topic, particularly for IT-enabled services. In one case a Pakistani subcontract worker threatened to post US patient medical data on the Internet if his financial claims were not met.[29]

- *Corruption risk.* This includes both grand and petty corruption. This risk applies to offshore subsidiaries, while firms outsourcing offshore are largely immune to this risk because the cost is absorbed by the offshore provider.
- *System security risk.* There is increased likelihood of insiders inserting malicious code or leaving open vulnerabilities, or entering corporate networks via privileged access. Terrorists are likely to exploit these routes in the years to come.

We move on to two project-related risks that are likelier when offshoring, or have greater severity of adverse outcomes, or both:

- *Contractual risks.* These are greater when offshoring, particularly when outsourcing. Adverse outcomes appear when there is a dispute between the parties which the parties cannot resolve. Foreign legal disputes may take longer to resolve, may be subject to corruption, or favor the local company over the foreign company. This risk can be mitigated by contracting with a company that has legal standing in your country (see more in Chapter 6 on legal issues).
- *Infrastructure risk.* The dependability of the communications infrastructure is lower in some offshore destinations. While the probability of failure may be higher, the severity is unlikely to be great. Companies mitigate this risk by securing multiple communication links to the offshore unit or provider.

The last of the offshore risks is at home: the *risk of societal and regulatory changes in your home country.* Societal changes impact company reputation and are difficult to anticipate. Political backlash due to offshoring began emerging in the early 2000s, in the US, UK, and Germany (see Chapter 12, Offshore Politics). The fear of a poor public image then begins to drive decision-making. For example, some American business managers have reacted by making offshoring a clandestine activity. They hide information about their offshoring as much as possible. Alternatively, they choose to outsource to a US-based provider with offshore IT resources, who then turns around and performs the work offshore. Additionally, companies may come under scrutiny by non-governmental organizations (NGOs) or the media, for perceived social or economic injustice such as "sweatshop" operations offshore. All of this reverberates into the organization itself. Employee morale may be damaged because of lay-offs or loss of career opportunities. The public relations backlash leads to bitterness among departing or remaining staff. This may lead to loss of key talent, primarily in high-tech firms.

Regulatory impacts to offshoring emerged in the US and Europe by the early 2000s. For example, the Committee of European Banking Supervisors proposed to ban outsourcing of "strategic or core activities." Some IT work may qualify within this definition.

The American Federal Deposit Insurance Corporation, which regulates banks, prohibits financial firms from hiring workers with criminal convictions; such a restriction applies to those managing systems in foreign locations. Finally, countries may become embargoed for other reasons in full or in part. For example, the US has, or has had, at least partial trade restrictions on a medley of countries in recent decades: South Africa, Libya, Iran, and Cuba, to name a few.

Assessing and managing offshore risk

Risk assessment needs to take place up-front, before going offshore, but also on an ongoing basis. Many firms conduct some type of country risk assessment before they enter a country (*ex ante*). However, it is rare that firms continue to conduct regular assessments once they already have operations in-country. This is a mistake. Companies need to continually assess risks by collecting data from experts, but particularly from people who are in-country, on the ground. The sensationalist news coverage that we all watch is not a reliable source for risk assessments.

The process of *risk assessment* is largely a qualitative exercise best done with multiple managers' input. The risks are listed, just as they are listed in Exhibit 2.2. Then the probability of each risk type occurring is assessed either by number (e.g. 0–10) or label (e.g. low, high). The consequences of each risk occurrence must then be assessed. For example, a system security break-in may be rated as having severe consequences. Finally, managers must review mitigation approaches for each of the risks, particularly the higher-probability and higher-severity risks. Managers can act to specifically lower the probability of an adverse outcome occurring and can lower the severity of the adverse outcome.

There are differences between large and small firms regarding offshore risks. Strictly speaking, for a large firm, the risk is the mathematical expected value, namely the product of the likelihood and the severity. However, for small firms, some risks, such as IP risk, can be so severe as to lead to the company's downfall. Is such a risk worth the cost savings of offshoring?

In closing, offshore cost savings and risks all tie back together. Executives weigh greater risks against greater costs savings. "[…] We are debating if the [offshore] cost savings are worth the IP security risk"[30] is an illustrative quote by the Vice President of US-firm New Health Science, which develops medical products to help detect circulatory abnormalities. Indeed, views of IP risks and their close cousin, loss of proprietary knowledge, vary greatly in the software community. In spite of the knowledge about the risks, as well as knowledge of known cases, most software product firms' managers are not concerned about IP risk.[31] Since managers are aware that there is a chance of the adverse event taking place (the probability is non-zero) then they may be assessing the severity of loss as being low. It is also possible that they are simply assessing the benefits to be greater than the severity of loss.

Concluding lessons

- Wages are only part of the story: if you hire directly, pay attention to benefits and fully burdened costs. If you outsource, pay attention to charge rates and onshore rates.
- Summarized country wage data represent only a first step: pay attention to the significant differences by region within country and to significant differences by position (junior programmer versus senior project manager).
- Know the "extra" offshore cost items, and then, most importantly, manage them closely.
- Benchmark your current processes properly so that you can make an informed decision about whether offshoring leads to real cost savings.
- Assess the trade-offs of the four major decision factors:
  - cost savings due to wage differentials,
  - "extra" offshore costs,
  - strategic benefits of offshoring,
  - additional risks from offshoring.
- Know the nine offshore risks as they apply to your company.
- Continually assess offshore risks: first at the outset of offshoring and then on an ongoing basis. Assess both probability and severity of each risk category.
- Actively manage risk mitigation: diversify to multiple sources and countries; set up contingency plans, backup sites, mirrored systems, and test the contingency plans. Carefully document all processes handled by offshore companies and offshore personnel.

3   Beginning the offshore journey

Floral Systems (an alias) is a medium-sized Dutch software company. Driven by a need to reduce costs, it decided that the next release of its software product was to be built in India. The Dutch project manager had just met a representative of a large Indian provider at an American IT fair and decided to sign a contract with that firm. Problems started to occur almost as soon as the project got under way. Floral Systems used a development platform called Progress, which is not widely known in India. The Dutch firm had overlooked the fact that the Indian staff had no experience with the latest version of Progress, which was quite different from earlier versions. Knowledge transfer from Floral Systems to the provider also proved difficult. One year later, the offshore project was a failure and was abandoned. The company not only wasted a lot of money, but because its next product release could not be delivered on time, clients started to lose confidence. The project manager had already lost his job.

A more successful Dutch example is Metatude, which was founded in 2000. At that time, due to the IT labor shortage, this start-up could not recruit experienced software engineers. It had no choice but to go offshore. Initially, Metatude investigated three countries: Bulgaria, India, and Bangladesh. Having prepared itself well, and weighing the trade-offs, it decided to focus on Bangladesh. Two managers visited this country for two weeks and had meetings with various service providers, foreign users, as well as the local Dutch embassy. Three of these providers were selected to bid on Metatude's request for proposal (RFP). Metatude chose one of these providers and then successfully built its very first software product offshore.

The offshore journey is not always easy, as the first Dutch example above illustrates. There are many companies sending out work to far away countries and experiencing problems, disappointments, and eventually pulling the plug. We often hear stories of failure. For some clients, India has even become an abbreviation for "I'll Never Do It Again".

While a few managers will be thrilled by the challenge of the offshore journey ("Yes, I want to go to exotic countries."), they are a minority. For most staff, stepping into the unknown is scary. It will bring them into contact with companies they have not heard of before, mostly from countries with which they are unfamiliar. On top of that, journalists have given exotic labels to offshoring, such as a "Passage to India", as if today's IT manager is going off to the land of the rajs in the 1800s.

Although still a novelty for many organizations, developing software offshore has a history of more than two decades. In the early period, before the Internet, it was difficult for the pioneers to find information, and clients had to rely on a telex, a fax machine, and on poor telephone connections for communication. It is a wonder that any international collaboration could take place at all. Today, there is an abundance of information, communication is easy, there are thousands of providers to choose from, and we can quickly learn from the lessons of many companies that have already offshored. Some of these lessons involve failures, though in general offshoring is not as difficult as some potential clients fear.

For example, the overhead of managing offshore projects is often overestimated among companies not using offshore services. The security risks and the nightmares of cultural and language differences are sometimes overstated. An A.T. Kearney study revealed that more than 80% of firms surveyed said that the quality of functions sent offshore was as good as or better than before.[1] Even many of the early adopters of offshoring that contended with less mature markets still succeeded most of the time.

There are, however, some common outsourcing mistakes, such as having overly optimistic expectations, a lack of internal support, having unclear specifications, or selecting the wrong provider. The principal error is underestimating the importance of preparation. Therefore, this chapter is devoted to preparation for *the offshore journey*. Journey preparation, depicted in Figure 3.1, consists of three major phases: laying the foundation, identification of potential service providers, and assessing and selecting the provider. Proper journey preparation will not guarantee 100% success, but will certainly diminish the possibility of failure.

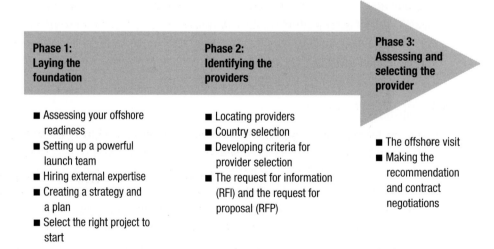

Figure 3.1 The three phases of preparation for the offshore journey.

Many of the tasks encompassed in the journey preparation are those of a general, well-managed provider selection process that may even be part of a procurement management methodology.[2] In other words, these tasks can also be used when searching for a domestic partner. However, in this chapter we emphasize the elements that are unique, or more difficult, or merit special attention when offshoring. Outsourcing is the most common way for organizations to begin the offshore journey. In fact, roughly 90% of the companies offshoring to India are outsourcing. Therefore, the focus in this chapter is on the offshore *outsourcing* journey.

A key question before journey preparation begins is: How long does it take? It took a large Dutch organization we know one-and-a-half years to conduct internal discussions, to select the first projects, and to finalize the search for suitable offshore partners. In addition, several trips abroad were needed to convince the managers of the quality of these offshore firms. While this is an extreme example, it illustrates our contention that a firm should not be rushed. We have helped firms move diligently through the three phases while making decisions quickly. Most small firms can do a respectable search in 2 months. Large companies, which intend to outsource large projects and want to assess different partners, often need more time. For these large firms, it is not uncommon if the provider selection process takes as much as 6 months or more.

Phase 1: Laying the foundation

Some users begin the offshore journey haphazardly, running off to sign a contract with the first provider they meet. This sometimes works, particularly with simple, small projects. In general, however, a deliberate approach based on a vision, with clear targets and with a proper provider selection process is required for success. The first phase in the preparation for the offshore journey is to lay a solid foundation. It includes the tasks of assessing the organization, setting up a launch team, creating a strategy and a plan, and selecting a (pilot) project. Some tasks in the phases are largely iterative with lots of "feedback loops" and need to be done in parallel; their order here is not always suggestive of a sequential ordering.

Assessing your offshore readiness

Is your organization really ready for offshore? For most clients, outsourcing software work to an offshore provider is not "business as usual" and several questions need to be answered first. Has there been offshore experience in the past? How is the maturity of the project management? Are they capable of managing international projects? How is the organizational flexibility? Is staff familiar with working with people from a different culture? Are they willing to travel? Will the employees of the organization accept this change in work norms? How is the complexity of the technical infrastructure? Are the processes stable and in place?

Companies need to consider taking the hardest step early on: improving internal processes first. These improvements include instituting mature project management practices and collecting accurate performance metrics for benchmarking. In a survey among North American offshore clients, 53% of the respondents reported having challenges in project management skills, 51% did not have good processes for specifying the work, and 48% did not have the right metrics for managing the performance of the provider.[3] Some of these companies were probably not ready for offshoring or, at best, should move forward cautiously.

When in doubt, the organization should conduct an internal assessment of its offshore readiness. As part of the offshore assessment, managers need to do a risk assessment and have an honest discussion about their tolerance for risk. This should take place during Phase 1 of the offshore journey preparation. If the outcome is negative, a no-go decision should be made. As a matter of fact, these negative decisions do happen. In such situations, organizations can consider offshoring at a later time. A Dutch company decided in favor of offshoring 5 years after a previous no-go decision was made.

The large Dutch bank ABN Amro is offshoring to several providers in India and Pakistan. It uses a spiderweb chart to assess the offshore readiness of its internal business units (see Figure 3.2). The chart is a useful focal point for decision-making dialog. Values are given on a scale from 1 to 5, for six internal criteria: the maturity of IT project management, the complexity of the technical infrastructure, the organizational flexibility, previous offshore experience, IT operations and support capacity, and the maintenance capacity.

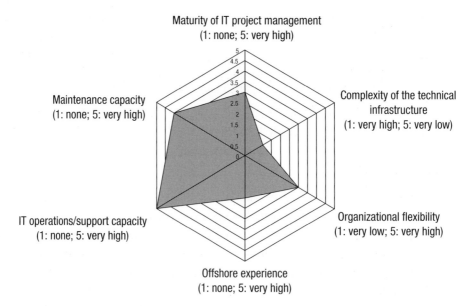

Figure 3.2 Applying the spiderweb chart to a business unit at ABN Amro Bank to assess offshore readiness.

The spiderweb chart shows internal strengths and weaknesses at a glance: the larger the shaded area, the better the chances to succeed offshore. Low values are warnings, and Figure 3.2 indicates that the high complexity of the business unit's technical infrastructure is a concern. On the other hand, it is strong in maintenance and project management, and it had some previous offshore experiences. Given these strengths, this business unit could consider offshoring, provided the weakness can be addressed. For example, it might be necessary for some of the offshore provider's staff to work onsite to master the complex technical infrastructure and to help the bank's unit adjust.

Setting up a powerful launch team

Since offshoring is more complex than domestic sourcing, it is critical to establish a powerful launch team that builds a strategic vision, demonstrates commitment, and transitions into implementation. The launch team is responsible for all tasks during the preparation of the offshore journey. This is the team that creates the plan, builds organizational support, selects the provider, and negotiates the contract.

The launch team should be small in order to be agile and to make quick decisions. Its members should include people who are open to offshoring, have an international outlook, and are willing to travel. The launch team may report to a steering committee, consisting of the chief information officer (CIO), senior IT executives, business unit managers, and legal staff. The launch team should also have good networks inside the organization in order to build internal interest and sponsorship.

The team members need to develop deep expertise in offshoring. Besides the many useful websites and research reports,[4] they need to seek knowledge both inside and outside the organization. There may be people inside the organization with offshoring or other international expertise that can be tapped for knowledge or brought in as supporters and sponsors. On the outside, team members can go to professional cocktails, visit seminars and trade fairs, and attend country-specific events. They can ask around in their own professional network and talk to other IT shops with offshore experience. The team should identify organizations similar to their own and find out about their experiences; it is not necessary to reinvent the wheel.

Hiring external expertise

Since many offshore countries and providers are not well-known, managers with little international experience should probably not embark on an offshore journey alone. Consider buying knowledge from experienced consultants in order to be a more informed buyer. Proper preparation for the offshore journey is also time-consuming. Hiring external expertise on a full- or part-time basis speeds up the process and saves valuable time for the launch team.

Before 2000, there were hardly any specialized offshore consultants. Now there are many to choose from.[5] Some consultants are specialized in one country (which is often

India); others have a more global reach. These companies advise on issues, such as creating an offshore strategy, identifying projects suitable for offshoring, selecting countries and providers, conducting due diligence, and arranging site visits. There are so many offshore advisors that we even know of companies which rate and verify the expertise of potential offshore consultants.[6]

Creating a strategy and a plan

The offshore journey should begin with an offshore strategy and an operational plan. "Failing to plan, is planning to fail", is an appropriate proverb for this stage of the journey.

The strategic goals of offshoring should be defined and understood at the outset of the journey. Of course, in most situations, the goal is simply to achieve cost reduction based on wage differentials. But, this is too vague a goal to be successful in the long run. A more specific goal could be: "Reduce the total IT budget by 7% within 3 years by using offshoring resources". Companies should also begin to formulate other offshore goals that are more than merely cost reduction goals. We discuss other strategic goals, such as an increase in speed and flexibility, later in Chapter 5.

Once the strategic goals are articulated, they need to be operationalized. Since organizations spend several years learning to manage the offshore cooperation successfully, an initial long-term plan should be created. The offshore plan should begin to articulate the future human resource (HR) needs; more specifically, the skill sets of your IT staff and the capabilities required from them in the future. The plan will state which projects are most suitable for offshoring (which we discuss in later in this section). It will prioritize all activities needing longer lead times (e.g. hardware and software requirements at the offshore site, additional training for offshore staff and internal staff, and knowledge transfer). The plan will also have an early risk assessment, which includes identifying risks and planning for mitigating each of these risks.

As part of the offshore plan, the launch team will draw up a budget for all three phases of the offshore preparation. The preparation costs can be substantial. Search costs include significant organizational time in documenting requirements, determining which projects or processes are to be offshored, and many hours on provider assessment. Search costs also include external costs, such as hiring external advisers, hiring legal counsel, travel costs, and translations.

In many organizations, launching an offshore initiative will involve preparation of a *business case*, which is the financial justification for the offshore strategy and consists of an estimate of the costs, the cost savings, and offshore risks. The business case is usually a document accompanied by a formal verbal presentation to management. Depending on the sequence of decision-making, the launch team may want to make a preliminary business case for offshoring at the very beginning of the preparation phase. Others will develop the business case later in the journey preparation stages, after the strategic vision is accepted, the plan is stable, and perhaps after making preliminary offshore visits.

The business case needs good data. A preliminary estimate of offshore costs should not overlook the "extra" offshore costs, such as additional overhead and knowledge transfer, which is the significant investment of time and effort in educating the offshore staff about the system and its uses. The other data are good internal performance indicators (which were mentioned as part of the assessment of offshore readiness). Performance indicators are needed to measure success. Organizations must know their real internal costs and productivity figures: on-time delivery, user satisfaction, service availability, response times, and error rates. If these are not quantified, then it will not be possible to judge offshore success or failure. The business case also benefits from good qualitative data: the offshore activities of competitors could serve as a benchmark. Survey the successes and failures of other organizations and draw lessons from both as they apply to your own company.

Finally, the offshore plan must anticipate a key organizational issue, namely that people never like change and that organizational resistance is one of the key barriers to offshoring. There are numerous *change management* approaches to address such resistance. They include implementing measure and reward systems to motivate offshoring, the creation of new organizational structures to support change, funding demonstration projects, and education about offshoring. These and other mechanisms of organizational change are described in Chapter 7. The HR department should be involved early, in part to ensure retention of critical people. HR should define a staff retention plan, arrange for retraining programs, plan for internal or external transfers, develop redundancy packages, and plan for dialog with trade unions.

Selecting the right project to start

The offshore journey can begin with a methodical, systematic, comprehensive inventory to decide on the right first project, or follow a more organic decision-making process and go with a *pilot* project. We describe each of these approaches here.

Careful preparation for the offshore journey means that the company inventories its applications and systems. Many questions need to be answered to assess if applications are suitable for offshoring. Is the scope of the work, and the functional requirements, clear? Is the technology being used standard or emerging? What is the maturity of the application? Are the connections with other applications tight or loosely coupled? What is the size of the work? Is knowledge transfer going to be difficult? Is there sufficient documentation? Can most of the work be done offshore or will it involve an onsite component? Is the implementation in a single location or multiple locations, or global?

From such guiding questions, the launch team will get various derived measures, such as criticality, complexity, annual costs, stability, and size. Applications and systems should be ranked as having a high, medium, or low potential for offshoring. The financial impacts of offshoring will also differ for each of these: they can give low, medium or high savings. Based on the offshore potential and the potential financial returns, the most suitable application(s) can be selected and a more detailed cost savings

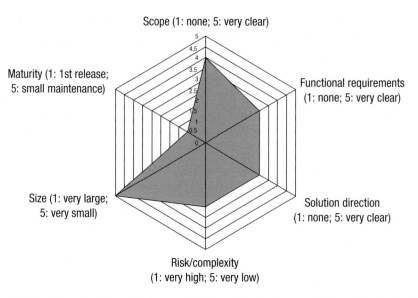

Figure 3.3 Applying the spiderweb chart at ABN Amro Bank to assess which project is suited for offshoring.

analysis can be done. This is the core of the offshore "business case". The summary findings will become part of the "scope definition document", which can be used to internally sell and communicate to management and staff.

Figure 3.3 shows another spiderweb chart used by ABN Amro Bank – this one is used to assess which project is best suited for offshoring. Values are given on a scale from 1 to 5, for six criteria (scope clearness, functional requirements, solution direction, risk and complexity, size, and maturity of the application). The spiderweb shows that the maturity of the application is a reason for concern since, in this case, it will be a first release. This requires specific attention: it might be necessary to hire specific offshore expertise, or to build a prototype first.

The Offshore Stage Model, introduced in Chapter 1, reveals that most firms move through an experimental phase as they begin their offshore journey. They start slowly and test the waters for a year or more. In the spirit of starting slowly, consider first offshoring your company's less critical projects and systems. These are small, structured, and non-strategic jobs, such as programming, testing, or conversion work. For example, you can start with support activities and move later to new development work. It is easier to offshore a migration project, where functionality is stable, and only the technology platform differs, than a new application to be built from scratch. Another approach is to start with tedious work, such as maintenance. This will foster internal support, since it gives employees the opportunity to work on more interesting tasks, such as new technologies, and to learn new skills.

If your company has no experience with offshoring, conducting a *pilot project* is recommended. A pilot project is useful to test if your company is ready for offshoring, but also to test a certain country, or to assess a new provider. It also allows you to set another, more advanced milestone, before the final go/no-go decision is made on offshoring. If the pilot proves satisfactory it can be used internally to communicate its successes and to learn from its problems.

A pilot project should be a low-risk task that is allowed to fail. It should have a relatively low level of complexity in an isolated environment (e.g. no interaction with legacy systems). Some companies select a project which they already did themselves in-house. A pilot must have the right size in order to be meaningful: one that is too small is useless; too large is not appropriate for a test. A pilot can run from several weeks to 6 months.

Measurement and tracking are needed to conduct a proper pilot. The pilot could contain intermediate products (detail designs, prototypes) to track performance. Overall performance should be measured using the following criteria:

- Quality, both technical (e.g. number of bugs) and functional (e.g. does it meet specifications).
- Cost and productivity (a provider might not have the lowest rates, but can be faster).
- Project cooperation, such as project style, communication, responsiveness, problem solving capabilities, documentation, and planning discipline.

Some clients have been quite creative and disciplined in inventing offshore pilot methods. We illustrate this with two clever cases. The first is a small American software firm, SSI, which attempted to conduct parallel pilot projects in India, Russia, and China in order to choose a country: the detailed case study appears in Chapter 4. The second is the case of Sogeti Netherlands, a large IT services provider, which wanted to test an offshore provider. In order to do so, Sogeti developed the same project twice: both in-house and by contract with the offshore provider:

Sogeti's goal was to decide if Indian provider NIIT would be a suitable partner. Sogeti decided to conduct a comparison and perform much of the project life cycle twice (from low-level design through integration): both in-house at the Dutch Sogeti office and offshore at NIIT in India.

Sogeti chose a real-life project that it was conducting for one of its clients: a web application involving document imaging and business rules supporting workflow, running on a .NET/Citrix platform. The pilot's size was 700 Function Points. One Sogeti staff member made a visit to India for knowledge transfer, though it was not required by the Indian team to make any onsite visits. A knowledge portal was created for communication between the two teams. This served as a single repository for technical documentation and deliveries, query resolution, and project management documentation. Processes were defined for project planning (e.g. weekly teleconferences), quality assurance, metrics management, requirements management, and delivery management.

The results of the pilot were favorable for the provider and for the offshore decision. After the pilot was complete, Sogeti assessed that 50% of the total work could be done offshore, resulting in a cost reduction for the client of 26%.

The quality in India was comparable with the quality of the work done in Holland. The functionality and the deployment were good in both cases. Finally, while the Dutch project team met all pilot project deadlines, the Indian project team was ahead of schedule most of the time.

Phase 2: Identifying the providers

The second and third phases of the offshore journey deal with identifying and selecting the offshore provider. This is a complex issue for most new clients, since they are not familiar with the global bazaar of providers. As we noted before, some clients select an offshore partner haphazardly without even going through an RFP stage. Since this increases the risk that you end up with a less appropriate partner, it makes business sense to spend sufficient time on these phases of the offshore journey.[7] There are several specific tasks that form Phases 2 and 3, and these require a plan, with a timeline and deliverables.

Locating providers

The supply of offshore providers is growing fast. In India alone, there are thousands of companies offering IT services. The largest companies are represented in the major markets and can be easily contacted for information. Clients in a small country, such as The Netherlands, can already choose from almost 100 locally represented offshore firms or their agents. Most of the IT associations in offshore countries have web-based resources with member company lists.

Local IT organizations can be helpful as well. An example in the UK is Intellect, the Information Technology Telecommunications and Electronics Association, with 1000 member companies. It has formed an offshore group that provides information about off-shoring, provider selection guidelines, provider lists, and a nice list of "do's and don'ts."

The provider market is also becoming heterogeneous: you might be able to request offshore services from your domestic IT provider. For smaller projects, consider the specialized online programming marketplaces that provide vendor lists or direct place-ment of projects for bid.[8]

Country selection

If the engagement is short (e.g. for one project, or a series of smaller projects), then choosing by provider is probably best. However, for any longer engagement, country selection should come first.

Fifteen years ago, the selection of a country was relatively easy, since there were only a handful to choose from. This is different today, with software producing nations in Central and Eastern Europe, the Middle East, Asia, and Latin America. The next chapter will describe the major location selection issues and presents 11 countries in more detailed sketches.

If general software skills are required, then the work can be done almost everywhere in the world. However, the supply in specialized skills (e.g. mainframe experience, or specific domain knowledge) is more limited. Factors to consider in country selection are: language issues, time differences, intellectual property rights protection, travel time, stability, and cultural differences. In case foreign personnel will have to work onsite for some time, arranging visa and work permits may be an issue (e.g. it is easier for Central Europeans to travel and work in Western Europe than for Indians). If you are seeking the lowest, rock-bottom price, it might be useful to leave the "beaten path" and explore alternative destinations: nations, such as China, Vietnam, or even North Korea.

Metatude, the Dutch company mentioned at the beginning of this chapter, examined three countries, including a nearshore option, before making the offshore decision. In many cases, however, the country choice is not the result of serious investigation, but personal contacts are the decisive factors. This is a desirable criterion for country selection. A survey on a group of American companies that outsourced to countries other than India showed that in most cases, the expertise of foreign staff working in the client company (e.g. from Pakistan, Indonesia, or Vietnam) was used in locating and selecting the offshore country.[9] We have also seen examples where a country was chosen because a manager was married to a local woman (from Romania, the Philippines, Thailand, and India), or because the director visited a country as a tourist (Nepal), or has religious or spiritual interests (Israel and India). In any case, it makes no sense to select an offshore location if your key people will dislike traveling to this country.

Foreign embassies often have information available on the local IT sector, and some countries have specialized trade promotion offices. Business tours are organized regularly by many nations eager to facilitate trade. These short trips, although designed for a general business audience, provide a broad perspective on opportunities in a certain country.

> *A visit is useful to gauge country fit. At the beginning of the offshore journey, one Dutch software company sent two managers to India for 1 week to meet several competent providers. Upon their return, they decided that the choice for India was wrong. One of these managers, who was designated to spend a long period abroad to lead the project, did not like India as a place to live. Eventually, the choice was made for Malaysia.*

Once the country is selected, a list of potential providers must be created. It is relatively easy to create a list of 10–20 companies for large countries, such as India, China, or Russia. This list might be smaller for other countries.

General criteria for selecting a provider:
- Company (size, growth rate, financial strength and stability, subsidiaries, and alliances).
- Human Resources (numbers, specialization, experience levels, education and training, and morale).
- Management (background and experience).
- Technical experience and vision.
- Functional expertise (business domain knowledge).
- Track record and clients (company reputation, references, and repeat clients).
- Processes (delivery processes, change management processes, and support processes).
- Methodologies (development methodology, project management, and knowledge management).
- Costs (offshore and onsite rates and additional costs).
- Quality initiatives and certifications (e.g. ISO 9001, CMM (Capability Maturity Model), P-CMM (People-CMM), Six Sigma).
- Ability to scale up operations.

Criteria that require extra care when offshoring:
- Infrastructure (telecommunications, IT infrastructure, and power supply).
- Software production environment (hardware, software, tools, and licenses).
- International experience.
- Language skills.
- Employee retention and turnover.
- Company culture (flexibility, hierarchy, responsiveness, hiring policies, and soft skills training).
- Cultural aspects and cultural awareness training.
- Global presence (local offices or a local representative).
- Legal issues (intellectual property protection, terms and conditions, and contract flexibility).
- Business continuity planning (backup systems, disaster recovery, and availability of alternative centers).
- Security (premises, physical access, and data privacy).
- 24-hour support and availability.

Exhibit 3.1 Key provider evaluation criteria.

Developing criteria for provider selection

Before any providers are evaluated, an organization needs to consider its key criteria for such an evaluation (see Exhibit 3.1). In the previous phase a suitable offshore project was selected and described in the "scope definition document". Based on this document, the criteria for provider selection and evaluation can be gathered and ranked. There are various hard (measurable) criteria, such as technical competence, experience, and costs. There are also soft elements (e.g. organizational culture and language) which are crucial but easily overlooked.

Naturally, the relevance of these criteria is different for every client organization. In most cases, the number of important criteria can be reduced to a smaller number, and six to eight criteria will often be sufficient. It might be useful to assign weights to these criteria, for example, by using a numerical value or a qualification (e.g. very relevant, relevant, or reasonable). Then a matrix can be created to assign scores for each of the candidate providers.

Some criteria are surprisingly unimportant to clients. Research among offshore users showed that factors such as process certification, or the size of provider, are of lesser importance.[10] Nevertheless, a small company should be careful about outsourcing to a very large provider. The small client will not get the attention that larger customers receive. In fact, the large Indian providers are known to turn away small clients and if they do accept contracts from smaller firms, they are less likely to assign their best developers to these projects. Conversely, a large client might not feel comfortable working with a small provider. A large Dutch company, which works with Indian Tier-1 firms, investigated nearshore opportunities in Central Europe. It decided to reject this option because the prospective providers were too small in size and would not be able to assign sufficient numbers of staff.

A survey among offshore users in Silicon Valley revealed interesting differences between IT end-user firms and software product producers regarding their provider preferences.[11] For the end-user firms, earlier experience in successfully carrying out large international projects was the number one factor, followed by cost. In the case of software product producers, the number one factor was prior technology experience. This was followed by the capability to provide services on an on-going basis, with the number three factor being cost.

The RFI and the RFP

While much information on offshore providers is available via websites or company brochures, this information is mostly of a very general nature and is not sufficient to assess if the provider can do the work according to your criteria.

To narrow down the provider list, solicit initial information, based on your specific questions, and send out requests for information (RFIs) to a number of companies. An RFI contains the following main sections:

- Introduction to your company.
- Basic project information (but without too many details).
- Questions about the provider (e.g. geographical locations, history, management, number of employees, turnover rate, processes, infrastructure, security).
- Questions on services offered (e.g. domain expertise, platforms, skills, number of experts, training level, customers, indication of tariffs, the use of subcontractors).
- Questions on strategy (e.g. vision, market share, partnerships, and alliances).

Use standard questionnaires with only your most pertinent questions; asking hundreds of questions is useless. Sending out RFIs to a very large number of companies is also a waste of time.

Experienced clients can bypass the RFI stage and expedite the provider selection process by narrowing down the field of candidates and selecting the best two or three; these will be approached with a request for proposal (RFP). Alternatively, limit the number of RFIs when personal recommendations are available.

After receipt of the RFIs, the responses will be evaluated using your firm's key criteria as taken from Exhibit 3.1. Only the most promising firms will be scrutinized in more detail, and they will be issued RFPs. The ones you did not select should receive information as to why they were not chosen. In general, the shortlist will consist of a small number of companies (e.g. two or three if a company needs one provider).

The RFP contains several sections:

- *Guidelines for responding to the RFP.* contact details, number of pages, format preferences, requirements to be addressed, and a deadline date.
- *Project-specific information.* This is based on your "scope definition document", and it must be sufficiently detailed for the provider to understand both the business issues and the technical issues of the project. It includes specifications, performance criteria, hardware and software requirements, communications requirements, skill requirements, training requirements, and documentation requirements.
- *Special questions.* You can ask the provider for creative solutions by using specific questions, which will bring out the provider's creativity, domain, or technical knowledge.
- *References.* A request for details of the provider's most closely related projects.

A sample contract may be included along with the RFP that includes issues regarding licenses, warranties, penalties, and incentives. You may also request a format for the financial proposal. One of the key financial parameters is T&M (time and materials) versus Fixed Price. T&M contracts are appropriate for projects that contain a great deal of uncertainty, are likely to change, or are complex. If the provider is asked to make a Fixed Price proposal the project should be stable and well defined. If this is not the case, then any provider bid will be useless (we have seen proposals from India where the highest bid was 30 times higher than the lowest!). However, even if the specifications are clear, there can be big differences.

The RFP process should not be made too complex. The RFPs are often lengthy documents asking too many questions that are not really useful. This is a waste of time and money for everyone involved.

Phase 3: Assessing and selecting the provider

The final phase of the offshore journey begins when the providers' RFP responses are received. The launch team will evaluate the responses and relate them to the selection criteria.

You must be aware of false promises and not rely solely on the RFP responses. Some providers, eager to get new clients, claim to possess skills and expertise which they do not have. Culture can complicate business discussions as well: some Asian salespeople are reluctant to answer "No" to questions of a foreigner, since this is considered to be rude.

Due diligence is critical in order to clarify and validate provider capabilities. Without such validation, weaknesses will not be discovered until at a later stage. Therefore, construct a due diligence plan to validate the information, with your company's selection criteria in mind. The following activities are advised to gain additional insight:

- *Call up references.* Contact the provider references and set up visits. Have people from the same level engage in these discussions. Ask the references about unforeseen costs they encountered in their offshore engagements and use the discussion to learn how to make the relationship work. Ask the provider to speak with clients that outsourced projects that had not gone totally smoothly.
- *Set up local meetings with providers to discuss their abilities and to address your criteria.* These meetings should be structured around a confirmed agenda and should include an overview of your operations and your offshore objectives.
- *Pay attention to the soft elements.* These are criteria that we emphasized previously: personal fit, and overall cultural compatibility, such as similarity of values, "feeling comfortable", "trust", and, in the figurative sense, "speaking the same language". Some companies will be easier to work with than others.

Some clients are bargain shoppers and are obsessed with the lowest possible rates. We stress that price should not be the dominant criterion; if it is, then most of the recommendations in this chapter (and this whole book) will have been ignored. One high-tech manufacturing company spent more than a million dollars evaluating and selecting an offshore service provider. Eight months into the implementation, issues emerged that escalated the program management costs. New providers (not considered in the initial evaluation process) had higher unit rates but were able to reduce the extra costs significantly.[12]

Figure 3.4 presents another spiderweb chart – this time for assessing offshore providers. Values are given on a scale from 1 to 5, for eight selection criteria: technical capability and experience, functional capability and experience, stability, infrastructure, historical relationships, flexibility, reputation (including track record), and productivity. This chart shows an area of concern: the provider's functional capabilities are somewhat limited. This might not be a problem if the provider is being assessed for a small maintenance project, but it will be risky if the project is a first release. In that case, knowledge transfer from the client to the provider will require specific attention and investment.

The offshore visit

The initial offshore visit has become a rite of passage on the offshore journey and has filled up many hotel rooms in India. Managers embark on a foreign trip to assess potential

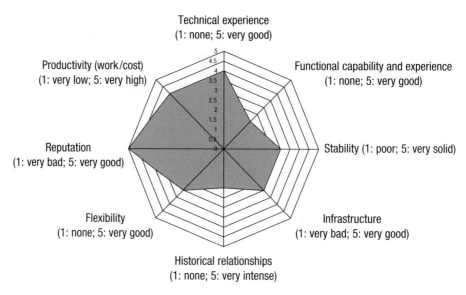

Figure 3.4 Applying the spiderweb chart at ABN Amro Bank to assess providers.

providers in detail and to get a feel for the country. Site visits abroad are recommended in case of large or complex projects, if long-term cooperation is required, or if the goal is that the provider build a great deal of specialized knowledge, which makes it expensive to switch providers.

Decision-makers have different preferences for the timing of the offshore visit. Some have taken the trip very early in the journey preparation, in order to educate themselves about offshoring and to get a feel for the offshore landscape. Others time their visit at a later point, after receiving the providers' RFI responses, in order to meet a group of promising companies. Still others wait until they have received concrete proposals from the most promising offshore providers. Usually, the offshore visit is conducted by the launch team members, who are at least somewhat supportive of offshoring, but we also know of delegations that took along managers that were reluctant to offshore in order to win them over. The reluctant managers can often be convinced when they see the offshore facilities with their own eyes and when they meet foreign staff in person.

Large providers will receive visitors at their headquarters, but this is not always the location where the actual work will be done; remote centers might have to be visited as well. Take the opportunity to "get a feel" for the country: visit the national software association; visit your embassy; talk to other users on similar trips; and get out of the offices and hotels, and do some sight-seeing.

Site trips can be exhausting and involve a great deal of time, effort and money, and will be disappointing if not properly planned. The objectives of the visit should be carefully defined upfront and not during the airplane flight. Some clients make the mistake of being overeager, and visit too many firms in too many places, of which most are not of any use. And if the actual visit consists merely of listening to sales pitches, then all providers will look the same.

The providers' standard onsite briefing consists of strategy presentations, an overview of capabilities, case studies (including demonstrations of software and life-cycle documentation), team meetings and a tour of the premises. You can request to speak to a project manager in order to get a detailed review of a past project. Also, seek out someone from quality assurance. Ask to review a project that did not go well, and ask why and what corrections were made. In addition to meeting with the provider's sales team, you should also meet with the technical staff and the support units, such as HR and training. A checklist should be used to tailor the meetings.

While the offshore visit can be very important at the beginning of the journey, it is not always necessary. Companies have offshored successfully without any of their employees having to make use of a passport. Generally these cases are in very small firms with small budgets, or for a small project, or if a local representative is available to act as a liaison.

Making the recommendation and contract negotiations

Many providers are good, but which one will be the best? Based on all available information, the final evaluation and selection can take place. Phase 3 ends with the final selection and the recommendation to senior management. The recommendation includes the project objective, the scope of functionality to be outsourced, the selection methodology, a candidate list, candidate functional and fee comparisons, and, of course, the nomination. The recommendation will include a financial justification. There is also a possibility that the launch team determines that offshore outsourcing is not a correct business decision, or that the selected providers, or the selected country, are not appropriate.

Once the provider is selected, the contract terms need to be finalized. Legal and contract issues are described in detail in Chapter 6, with a few key items noted here:

- *Pricing and additional costs.* The key issue here is to have a full understanding of which costs will be covered by the provider for items, such as travel to the client site, software licenses, training, and hardware acquisitions.
- *Issues of intellectual property, warranties, and confidentiality.*
- *Incentives and penalties.* These are the mechanisms by which the provider's performance is better aligned with the customer's goals.

In general, the contract is secondary to building a relationship. Long and overly detailed contracts are often an indication that there is little trust and that relationships were not built during communications with the provider. Such cases do not foretell a successful offshore project. Understand that the provider also has constraints and needs, such as the set-up time for the infrastructure and the project team, investments in training, and investment recovery. An environment must be created where the business interests of the two parties are in alignment over the life of the relationship. Relationships may also be important with those firms that were not selected; contact them and explain the reasons why they were not selected, since you may want to work with them in the future.

You are now ready to embark on the offshore journey. Bon Voyage!

Concluding lessons

- The principal mistake, as companies begin their offshore journey, is to underestimate the importance of careful preparation. Sufficient time and budget must be available for laying the foundation, provider identification, and the final selection.
- Create a small, agile but powerful project launch team with members who are open to offshoring. If you cannot do this with internal resources, consider buying knowledge from experienced consultants.
- Internal resistance is the greatest barrier to offshoring. Visions and benefits must be communicated clearly early in the journey. Consider change management approaches to address internal resistance.
- A low-risk pilot project is recommended. It can be used to test if your company is ready for offshoring, to test a certain country, or to assess a new provider.
- Be aware that the sequence of country-first or provider-first is dependent on several key factors. If the engagement is short, then choosing by provider is probably best. For a longer engagement, country selection should come first. The country choice can be the result of investigation, but it is legitimate to make the choice because of personal contacts or personal reasons. Select a location your key people will enjoy traveling to.
- The criteria for provider selection criteria are different for every client organization. Be selective in sending out RFIs and RFPs to potential providers; long lists are a waste of time.
- Do not rely solely on the RFP responses. Due diligence, the careful investigation of the provider, is critical. Without such validation, the provider's weaknesses will not be discovered until a later stage. A site visit abroad is recommended in case of large or complex projects or if long-term collaboration is desired.
- The contract is secondary to building a partnering relationship. Try to create an environment where the business interests of the two parties are in alignment over the life of the relationship.

4   The offshore country menu

Close to 100 nations are now exporting software services and products. The "offshore menu" is immense, with many nations to satisfy any taste. These nations span the economic spectrum from newly industrialized economies, through transition economies, to developing economies, and even some least-developed nations.

The 1990s was the first decade of software's true globalization.[1] The 1990s saw the rise of three celebrated success cases, the "three 'I's" – India, Ireland, and Israel. These were the three nations that seemed to appear overnight as global centers of important software activities. One of the most interesting features of the "three 'I's" is that each of these three nations developed and specialized in different aspects of software. That is, each one progressed to become a global software player in different ways: India in offshore programming, Israel as an incubator of software products, and Ireland in programming services and localization services.

Three tiers of software exporting countries

The G7 nations[2] produced much of the world's software in the first few decades of the computer era. High-tech exporting used to "belong" to these nations with the USA as the hegemonic power in software. Until roughly 1990, very few nations exported software products or software services at any non-trivial levels, including that which today we call "outsourcing". The G7 nations are still at the core of the Tier-1 software nations (see Table 4.1), the *Mature Software Exporting Nations*. These G7 nations have a tradition of exporting high-technology and knowledge-intensive products and services. In particular, the USA (with its giants, Microsoft and IBM) continues to dominate world markets. The other G7 nations, Japan, Great Britain, Germany, France, and Canada, have had successful software (and computer hardware) industries spanning many decades. The one outlier among the G7 is Italy, which has never developed a strong software sector for an economy of its size. To the Tier-1 software nations we add several other advanced industrialized nations: The Netherlands, Sweden, and Finland, which have all had strong software export sectors.[3]

To Tier-1 we also need to add the "three 'I's" – India, Ireland, and Israel. All three nations have developed robust software export industries (we return to the three "I"s

Table 4.1 The 3-tier taxonomy of the world's roughly 100 software exporting nations[4]

Tier-1	Mature software exporting nations	Mostly industrialized nations such as: USA, Canada, UK, Germany, France, Belgium, The Netherlands, Sweden, Finland, Japan, and Switzerland Entrants from the 1990s: Ireland, Israel, and India. Entrants from the 2000s: China and Russia
Tier-2	Emerging software exporting nations	Brazil, Costa Rica, Mexico, The Philippines, Malaysia, Sri Lanka, South Korea, Pakistan, Ukraine, many other Eastern European countries, and several more elsewhere
Tier-3	Infant stage software exporting nations	Cuba, El Salvador, Jordan, Egypt, Bangladesh, Indonesia, Vietnam, and 10–20 others
Non-competing	Non-competing	About 100 of the mostly, small, least-developed countries of the world, including most African, and many Middle-Eastern nations. These nations have *few to no* software exporting firms

later in this chapter). Finally, we add the two newest entrants to the Tier-1 nations: China and Russia. As recently as 1999 it would have seemed far-fetched to classify China's software export industry in Tier-1. But China's software industry has been maturing so quickly in the early 2000s that this is no longer debatable. Russia's place in Tier-1 may still be marginal at this point.

The software exporting nations are classified into tiers based on three criteria: industry maturity, clustering, and export revenues. These criteria are soft criteria, and the tiers themselves are constantly changing. They will surely be defined differently a decade from now.

- *Industry maturity* connotes the nation's tradition of exporting software. Most Tier-1 nations have been exporting software since well before 1990, with Russia and China as the only exceptions. Tier-2 nations have been exporting since at least the mid 1990s.
- *Clustering* connotes some critical mass of software enterprises participating in the software export industry. Tier-1 nations have hundreds, and in some cases thousands, of firms exporting software products and services. Clustering also connotes a maturing collection (agglomeration) of secondary services to support software companies, including consultancies.
- *Export revenues*[5] are the magnitude of national software exports. Some representative numbers: India 12.5 billion USD (2004), Israel 2.7 billion USD (2003), Brazil 200 million USD (2001), Vietnam 30 million USD (2003), and Indonesia 30 million USD (2000).

Tier-2 nations are the *Emerging Software Exporting Nations*. All of these nations already have significant software export industries, exporting at 25–200 million USD per annum. Most of these nations have clusters of technology firms either in major metropolitan areas or in designated technology parks. These nations have dozens of organizations exporting software, but usually less than 100. We use the neutral term *organizations* because the unit that is exporting software may be a software subsidiary of a multinational enterprise, or a home-grown, independent software company.

Most of the Tier-2 nations are unlikely to move up and join the mature Tier-1 software nations. Their first liability is a small population base, which restricts their ability to grow large industries. A second liability is unfavorable conditions, such as political instability or immature stage of economic development. The strongest of these Tier-2 software nations, Brazil, Mexico, South Korea, and the Philippines, may coalesce and form an intermediate second tier within a few years, and separate and distinguish themselves from the smaller less robust countries in this tier. These more vibrant software nations are the larger nations that possess the wealth and large labor pool of educated human capital that is needed for growth.

Tier-3 nations are *Infant Stage Software Exporting Nations* with an insignificant impact on the global software market. Some Tier-3 nations have benefited from some foreign direct investment (FDI) in a number of their firms, but it has been small, and isolated to just one or a few firms. India's remarkable success is well-known, and governments in a number of these nations have woken up to the potential economic benefits of software exports. They are working to encourage the software export sector. This is discussed in greater detail in Chapter 10.

The software industries in Tier-3 nations are mostly "cottage industries," where companies are small and management is not professionalized. Transforming these industries will take years. Most Tier-3 nations will not move up to Tier-2 because of their relatively small size, which restricts their ability to grow large industries. More significantly, due to their stage of economic development, or political instability, the industry growth will be stunted.

What country to choose?

In this section we look at the factors that determine where companies locate their software work. After we cover a general list of criteria, two specific locational factors are discussed in greater detail: risk and government incentives. This is followed by a case study of a small American firm that was faced with a country choice decision and took an interesting approach to solve it: trying several countries.

The many factors to consider in location decisions

What are the factors that should be in your company's location decision? We begin with four high-level factors.[6]

1 *The type of activity that is going abroad.* That is, whether the activity is basic research, applied research, or development. For basic research the sites should be

located next to major national engineering universities, as most talented people are likely to be there. For basic research location costs and wages are less important.

2 *Duration of engagement*. Managers beginning the offshore journey sometimes ask: "what comes first, choosing the country or choosing the provider?" If the engagement is short – if it is one project (out-tasking) – then choosing by provider first is probably best. However, for a longer engagement, and certainly for any kind of captive center (subsidiary), country choice should come first. The company has to gain a deep understanding of the nation's context in order to choose the right fit.

3 *The firm's geographic orientation*. That is, whether the firm has regional or global preferences, or some other affinity. The Swedes called geographic orientation "psychic distance" to explain where their companies choose to locate their research and development (R&D) centers abroad. Americans call this "cultural distance." Geographic orientation explains why the British locate in India, the Spaniards in Latin America, and so on. One research study even measured the cultural distance between countries using a composite measure of cultural characteristics[7] in order to determine whether culture has an impact on foreign investment – it did.

4 *Motivation*. This refers to whether the firm's motivations are supply- or demand-driven. Supply-driven means that the firm wants to have access to resources such as high-end labor, low-cost labor, or technology know-how.[8] Of course, access to resources has been the dominant consideration for offshoring. Companies want access to cheap and vast pools of labor. Demand-driven means the firm wants *access* to local (offshore) markets. For example, China has drawn nearly every major high-technology firm to its cities for this reason. Companies gain proximity to important customers and markets – and in the case of China, easier access through governmental controlled markets. The other element of demand-driven motivation leads companies to locate where they can better redesign products for local use – localization.

The demand for resources often leads companies to locate in geographic "clusters." For example, of all foreign investment in India, Bangalore alone took more than one-third (38% in 2002).[9] At first, it seems counter-intuitive that foreign companies all come in from afar to tap resources and cluster next to one another, sometimes setting up shop in the same office park.

Yet, there are many good reasons to cluster together. Accessing the labor pool is easier in clusters: the top engineers and scientists are more likely to be in the cluster or move there. Companies want to be next to the best recruiting sources, and so they locate next to scientific competence centers, universities, or research institutes. Of course, this also affords them proximity to leading research activities. Furthermore, companies want to locate where, as one manager said to us: there needs to be a "buzz." Being close to major competitors is actually a good thing because it allows the firm to gather soft information about what competitors are doing. Finally, of necessity, in developing nations with their poor infrastructures, one cannot find support services outside the technology clusters.

Missing from the above list of four factors is *country specialization.* Why? Shouldn't technology executives make location decisions based on special national expertise in a desired field? After all, one thinks of France for wine, Japan for cars, and Switzerland for chocolate. In all these nations there are clusters of firms that specialize in producing or supporting each of their specializations. Are there no national specializations within software? The short answer is that there are no nations in Tiers 2 or 3 that have country software specializations. All of these countries have smart people, but they are offering generic skills. These nations compete with one another via other national factors, not software specializations. In some nations there may be one or two firms that have world-class capability in one dimension, but this does not constitute a national specialization. If a company is searching for very specialized expertise, it usually exists offshore, but rarely as a national distinction.

Let us take a closer look at the supply-driven factors, as these are most important for offshoring software in most cases. A.T. Kearney, an American consulting company, compiled an offshore location attractiveness index summarized in Figure 4.1. The index is made up of three criteria groupings to rate and rank nations qualitatively for offshore IT work:

- Costs, which includes wages, infrastructure costs, tax, and regulation costs.
- Labor, which includes business experience, labor force availability, education and language, and attrition.
- Business environment, which includes investor rankings, country infrastructure, cultural adaptability, and intellectual property (IP) protection.

Notice that weights in Figure 4.1 were chosen by the study's authors, while the weights for your firm are likely to be different. For example, cost may be weighted more heavily at your firm. In fact, the country decision-maker should see in these data a classic trade-off: those countries with fewer points on costs, such as Canada and Australia, have many points on "Business environment." And, vice versa, those nations with lower costs tend to have fewer points on "Business environment."

The "surprises" on this 2004 index, as the authors note, are the nations ranked immediately after India and China, namely, Malaysia, the Czech Republic, and Singapore. Malaysia, a country of 22 million people, is most noted for the massive government project in the Multimedia Super Corridor stretching from the capital of Kuala Lumpur to the new airport south of it. This geographic cluster includes the smart cities of Cyberjaya and Putrajaya, with their excellent infrastructure in connectivity and facilities. Malaysia's potential may well lie in IT-enabled services (ITES) as much as software. And due to its largely Moslem composition, the nation is becoming a preferred destination for the many nations of the Islamic world.

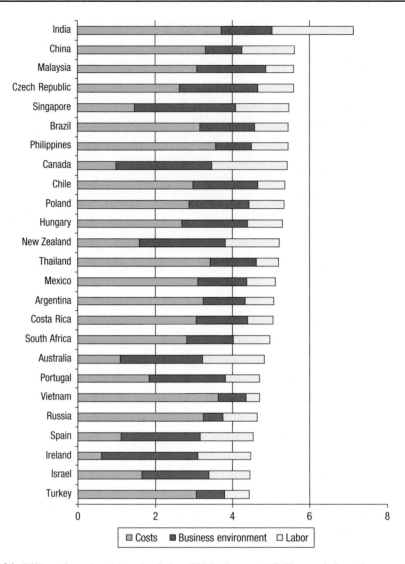

Figure 4.1 Offshore Location Attractive Index (2004). *Source*: A.T. Kearney's Location Attractiveness Index.[10]

Country risk

One of the location criteria is *country risk* (introduced in Chapter 2). Offshoring is about sourcing mostly from developing and emerging countries, which have historically been more volatile, less stable, less predictable, and less transparent. When offshoring, companies are exposed to increased risks of war, terrorism, rioting, uprising, confiscation, expropriation, and currency crises. Thus, the consequences of country risk can be severe because they affect, among other issues, *business continuity*, which is the ability of the firm to continue its core operations. Country risk is an important, sometimes dominant, issue. Larger firms diversify country risk by setting up operations and back-up sites in a number of countries, rather than just one location. Thus, in event of a crisis, they can shift operations.

There have been several tremors since the offshoring era began, such as the near war between India and Pakistan in June 2002, the September 11th 2001 attack in the US, and the period leading up to the American invasion of Iraq in early 2003. All of these tremors heightened uncertainty, resulting in some delays in decision-making and some project delays. But none of these tremors have seriously impacted offshoring. If a limited Indian–Pakistani war were to break out, it would disrupt a company's business by delaying some work and hindering travel, while a more severe war would have long-term consequences on a company's continuity if key systems and processes are in India. Note that war is considered a "force majeure" event, and in the event of war, if the provider fails to meet commitments, it is not considered as breach of contract.

The risk of conflict may well be overstated for many offshoring decisions.[11] First, critical production systems are usually not offshored. Second, software work tends to be somewhat isolated from the rest of the economy and needs little in the way of supply. India's software industry has seen only insignificant disruptions with all its tensions with Pakistan. Similarly, Israel's large software technology sector has not seen work disrupted by the nation's ongoing problems as a result of the Palestinian Intifadah.

Even Yugoslavia, during its 1999 war, as NATO jets bombed, saw software operations continue. When the war broke out, Clockwork, an Amsterdam-based Internet services company, had just set up an office in Novi Sad, Yugoslavia's second largest city. Several strategic targets in the city were destroyed during the attacks, such as its TV station, the oil refinery, and the bridge crossing the Danube. The residential area where the office was located was not a target, and the team of five programmers was able to continue the work. Broadband Internet connection continued to be available and communication with Amsterdam was not disrupted.

Country risk has a number of secondary implications. One of these is the risk *of government regulatory changes* (also called sovereign risk). A generation ago it was common to fear confiscation, nationalization, and expropriation. The Indian government, in 1977, actually made business so uncomfortable for IBM at the time that IBM pulled out of India entirely. It is hard to imagine these kind of events in today's global mood, but they may still happen. Less acutely, governments may change tax or subsidy structures that once favored offshore operations, making them less attractive. Thus, favorable tax treatments in, say, the Philippines or China may change. Governments may also change the regulations that govern technology joint ventures in order to favor the local partner.

Incentives provided by national governments

Countries compete with one another to attract foreign investors into their countries, states, provinces, regions, and cities. The competition for such Foreign Direct Investment (FDI) has become more intense in recent decades, with no signs of letting up. Companies that want to build software centers are especially welcome (see Chapter 10).

Governments induce FDI in software activities through three principle incentive mechanisms:[12]

- Governments reduce or eliminate various taxes. This can be done through a tax holiday (such as a 5-year tax exemption), or by import-duty exemptions.
- Governments subsidize certain activities and investments. Some investors will receive grants, although these are usually negotiated. A more common approach is a de facto rent subsidy in a technology park.
- Governments ease the bureaucratic process. This is done by giving the investing company specialized attention, such as one-stop investment and business registration processing, and simplified export and import procedures.

Some of these incentives are dependent on specific geography, specifically in technology parks or in tax-free zones, while other incentives are not specific to location.

Countries that rank in the top of "R&D subsidies and tax credits" are, in order: Israel, Singapore, Taiwan, Canada, and Ireland.[13] Not surprisingly, none of these most inviting nations are low-wage nations that compete largely on labor arbitrage.

Some examples of the range of government incentives appear in Exhibit 4.1. These incentives should not determine your company's investment choice, but rather they help tip the scales after narrowing country choices to the top two or three choices.

- Costa Rica – 8-year income tax exemption, followed by 4 more years at half of the standard 30% tax on profits; duty-free imports; unfettered capital repatriation.
- India – Software profits are not taxed until 2009, 10-year sales tax exemption, 5-year tax exemption on imported capital goods, special concession for power generation.
- Ireland – Cash incentives of up to 45% of the investment. Standard 10% corporate tax rate.
- Israel – Cash grant of 24% of investment, or a 10-year income tax break.
- Malaysia – 10-year tax holiday on profits and tax exemption on imported capital goods.
- Mexico – 30% of R&D expenses are tax free.
- Philippines – Software firms reside or register in high-tech zones (which can be in Manila or elsewhere). Once the firm is classified as an export-oriented firm it receives a 4-year tax holiday and duty-free imports.
- South Africa – Foreign software firms that conform to the "small and medium size enterprise development program" receive cash grants on a sliding scale, for example the first R5 million (about 500,000 USD) are matched up to 10%.

Exhibit 4.1 A sample of government benefits and incentives for foreign technology companies making investments.[14]

Case study Sport Systems Inc. shops creatively for an offshore provider

"From my previous experience with offshoring to India and to Vietnam I learned that there were not huge differences between the providers in each country – but, rather, that each country was very different: how programmers solved problems in each country was quite different." (Charles Angler, of Sports Systems Inc. (SSI), on why he decided to test providers from three different countries.)

This is an actual case. At the request of SSI all names are disguised.

Once SSI decided to develop its new software product offshore, it came up with an unusual plan. The company chose three offshore providers, one each from India, Russia, and China, to compete for the project. SSI planned to contract with each provider to complete an *identical* fixed-price pilot project. Comparing the results, SSI would choose the best performing provider for the full project.

SSI is a 20-person American software product company specializing in a distinct segment of the sports business. While the firm has been successful, it competes with many other firms in this segment. Charles Angler joined SSI in 2003 as Director of Systems Architecture. In his previous job he gained first-hand experience in offshoring. Some of his projects succeeded and others failed. Angler talked about his lessons:

"When you're in the negotiations phase, and you talk to these offshore companies, which you have never met, they'll all say they can do it, they'll say we have all these wonderful references, here's some sample code – and it all looks good ... but you want to look at the substance ..."

It was clear to SSI and to Angler that, given SSI's small size and limited funds, it could not develop its new product inside the company or pay the high rates that an American firm would ask. Angler identified and began negotiating contracts with three providers: Powercode Plus of India, TQ-Link from Russia, and Sun3 of China.

"The key success factor to offshore work is to create strong process – in particular, strong specifications using unified modeling language (UML)[15] and make sure that the offshore provider is experienced in iterating with UML," said Angler. Together with one of his business analysts Angler carefully specified the project and divided it into five releases. The pilot was made up of a subset of "use cases" that span the full process lifecycle. The pilot included a project plan, a graphical user interface, a UML-based logical view, a design for integration with Web Services, and a final .NET prototype (including test cases and results), with corresponding iteration plans throughout.

The SSI offshore contest was now ready to begin.

A problem immediately arose in contract negotiations. The Indian company, Powercode Plus, became too fussy about how it would treat the software code in case of dispute between the parties. SSI, a small player in a fiercely competitive segment, was very concerned about protecting its intellectual property, leading it to abandon the negotiation and eliminate Powercode from contention.

The Chinese company, Sun3, was very aggressive from the start, completing the project in 14 days. "They were working all hours," said Angler. "I knew because we required the offshore team to be on instant messager (IM) all the time – and they were almost always there, even with the enormous time zone differences." In fact, Angler was concerned that they worked too hard and would not be able to sustain this type of effort over a longer, bigger project.

Angler and others at SSI reviewed Sun3's code and concluded that it demonstrated reasonable skills in UML, although the offshore team was not familiar with some best C# coding practices. Sun3 programmers had created some object classes that were too broad and might lead to extensibility problems down the road. But, once SSI explained the deficiencies, the Sun3 team quickly brought the source code in line with standards. Sun3 completed the pilot for 2000 USD, using 150 person-hours, based on a charge rate of 11 USD per hour. "The interesting thing," said Angler, "is that during the entire project duration, we never once had a voice conversation with anyone at Sun3 – everything was through e-mail and IM."

Meanwhile, the Russian company, TQ-Link, took 1 month to complete the project. "We were very impressed with their work," said Angler, who noted that TQ-Link's work came in solid the first time around. TQ-Link demonstrated deeper knowledge of UML, produced a somewhat better user interface, and built a rigorous test suite. "And, I had excellent communications with their Boston-based representative, Sergey, who speaks excellent English." TQ-Link completed the project for the fixed price of 8000 USD, using 425 person-hours, based on a blended charge rate of 20 USD per hour.

"We were quite happy with both companies" concluded Angler. But, although TQ-Link was superior on several dimensions, the differences between the two firms were small. This magnified the cost difference, since the Sun3 rates were about half as much as TQ-Link. SSI chose the Chinese firm, Sun3. Angler added: "I suspect that the actual hours, for Sun3, was probably closer to 250 hours than 150."

Case lessons

Replicating several small pilot projects is an exemplary approach for choosing both countries and providers. It is worth considering for both first-time offshorers and experienced offshore managers like Angler. SSI's careful specification reduced mis-communication and focused the evaluation on the provider's true technical and business capabilities. On the other hand, SSI's decision was not as clear-cut. It was faced with two good choices. It seemed to choose cost over quality. Not all firms should make a similar choice.

Country sketches: the Big Three and eight more

We would like to give sketches of the 100 nations that export software. For practical purposes, we chose eleven nations that represent something of a cross-sample. We begin with the Big Three: India, China, and Russia; then cover the two other "I" countries briefly, Israel and Ireland. We then introduce six relatively unknown offshore destinations: Latvia, Romania, Malta, Vietnam, Bangladesh, and Costa Rica.

The country sketches begin with the "Big Three" offshore destinations: China, India, and Russia. Much of the volume of software offshoring is going to these three large nations. These nations are also *big* in the traditional sense, in that all three have large populations. In addition, these large populations also have a large, well-educated labor force in Science and Technology (S&T). In current industry parlance, we say that these nations have "scale" and a "deep labor pool." All three have seen substantial government investments in human capital (literacy, schools, and universities).

There are a number of other important common denominators among the Big Three. Of course, all three are already active in offshore software work, and in all three wages are low relative to industrialized nations. All three have opened up their economies considerably since 1990. Yet, all three nations are deeply divided economically, with computer and Internet penetrations at relatively low per-capita levels. All three have software organizations that are very rapidly improving processes and software quality. All three have seen most of the innovative software work take place inside the captive centers of foreign technology companies, rather than their own home-grown firms. All three have had little commercial success with independent indigenous product R&D. All three have poor protection of IP.

Broadly, India has become successful as a software factory, perfecting software production and delivery systems. Russia has had some success in algorithm-oriented software that requires invention and resourcefulness – generally called software R&D. With China's software export industry so young, many have been wondering if it will be factory- or algorithm-oriented. It is not yet clear where China will turn. China's success in manufacturing suggests that it will be successful in factory orientation. Chinese temperament suggests that it will be more successful in algorithms. It is quite possible that China will succeed in both.

We begin the country sketches with the Big Three nations.

India

India's success in software has been so extraordinary that it has been the subject of many books, and countless magazine and newspaper articles trying to make sense of how this poverty-stricken country became a glitzy, dynamic, high-technology powerhouse that makes European executives and Washington policymakers fearful.

India's remarkable success is driven by multiple factors. First, and most important, is its vast human capital that is well educated and English speaking. The national network of quality universities has been one of its gems. The elite Indian Institute of Technology, of which there are several campuses, accepts 3500 of 178,000 applicants a year, a selectivity rate of 2%. The scale of human capital is enormous: by some estimates, there are more IT engineers in Bangalore than in Silicon Valley (150,000 versus 120,000).

India's industry has successfully created a top layer of large, dynamic, multinational firms. Not just one or two successful large firms, but several. This is a feat that the Russians, so far, have failed at entirely, while the Chinese industry is still too young to judge. The top Indian firms (usually labeled Tier-1 firms, introduced in Chapter 1) are extraordinary success stories in their own right. By 2004 there were three Indian IT firms with more than 20,000 employees and two more above 10,000. The top Indian firms have also been unusually profitable, with gross margins of 45% and net margins of 30%, all while growing at 32% between 1999 and 2004. The firms' success in IT services is epitomized by capturing business with more than half of the largest US corporations (See Figure 4.2).

India has also been successful at creating synergies between similar knowledge-based sectors: it began with software services and used these competencies to move into software R&D and ITES. The ITES market has grown from near zero in 1998 to 2.4 billion USD by 2003, and is forecasted to grow eight-fold within 5 years. Indian firms have also been quite successful in software R&D, estimated at 1.3 billion USD in 2003, with 77 global firms having established direct R&D subsidiaries in India.[16] India has also grown a successful cluster of independent software R&D contracting firms ("labs for hire") that perform projects for foreign companies. The largest of these is Wipro, with 6500 engineers in its software R&D division.

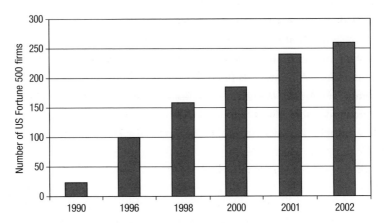

Figure 4.2 Number of US Fortune 500 corporations with some offshore IT work conducted in India.
Source: Authors based on NASSCOM data.

In summary, the story of offshoring is, in many ways, the story of success of the Indian industry.

China

The Chinese software industry was not on anyone's radar screen as recently as 1999. Like the modern Chinese cities that seem to sprout up almost overnight, China's software industry has emerged to become a global player in just 5 years. In China, more than any other nation today, structural changes happen quickly.

Chinese human capital indicators are as impressive in quality and quantity as those of India, with the exception of language skills (although there has been improvement in recent years). Chinese universities produce roughly 300,000 engineering graduates per year. While science and technology in universities has traditionally been too theoretical, this too has been changing rapidly. Approximately 30–45 universities have launched new, specialized schools in software. Such rapid adjustments show the ability of the government to redirect resources on a massive scale. China's human capital has been augmented by a reverse brain drain. Approximately 160,000 Chinese have returned with foreign education. Many are starting firms or choosing to work for foreign multinational firms. These returnees are bringing with them considerable know-how in technology, managerial project and process experience, experience in western business, and fluency in English.

The Chinese software industry is not as distinct as the Indian industry in two respects. First, the Indian industry does relatively little work domestically, while the Chinese industry does a great deal. Second, the Chinese industry is more closely tied to computer hardware and other manufacturing industries. China's software strengths have specifically been in embedded software at the interface between hardware and software: in telecommunications equipment, data communications, and wireless. Yet growth has taken place in all major software segments: in services, product R&D and embedded software, as well as in the related ITES.

The volume of exports of software products and services for 2003 was somewhere under 1 billion USD, representing roughly 10% of the Indian software powerhouse, but expected to reach 30% of the Indian volume by 2008.

Most large technology firms from the US, Japan, Europe, and India have software centers in China. Unlike India, in which the USA and Europe are the dominant investors and clients, Japan is a key investor and client in China.[17] The majority of large American technology firms, including Cisco, Intel, Microsoft, and Motorola, do some R&D work in China, including some innovative R&D. In total, R&D centers have risen from 150 in 2002 to 400 by 2004.[18] While Chinese software units have not attained the large number of world-class quality marks (Capability Maturity Model (CMM)) that Indian firms have attained, this is less significant for R&D activities.

Of the 2000+ indigenous Chinese IT firms, only a small number export software services or products. And none of China's pure software firms have attained the size of the major Indian firms. Noteworthy in size is Huawei, at the hardware–software interface. This is perhaps the most interesting Chinese player, as a major competitor to Cisco, with 2002 revenues of 2.7 billion USD (with almost one-third derived from exports). Most exports are to developing/emerging markets rather than the rich markets of Europe and America. Huawei has built a network of R&D centers outside China in India, Texas, California, Sweden, and Hong Kong.[19]

The Chinese software industry benefits from four complimentary growth drivers[20] which will all interplay in its offshore software industry in the coming years. The first driver is government-led development. Government support is quite strong: through procurement policies and through its influence at the national and local level. In China this makes an impact quickly. The second driver is to follow the India model by providing IT services to foreign firms. The third is to continue to attract Foreign Direct Investment driven by the desire for market access by global firms.[21] The fourth and last driver is to continue to be nourished by *brain circulation* – the return of thousands of Chinese bringing with them critical know-how.

Russia

The Russian software sector is by far the smallest of the Big Three nations. After the break-up of the Soviet Union Russia was viewed as having enormous potential in software but has not lived up to those lofty expectations.

Russia's strength is in its large workforce educated in science, mathematics, and engineering, including many with advanced degrees. Figure 4.3 points to the strength in Russia in the proportion of advanced degrees versus India. This strength should continue to manifest itself in software R&D. While the educational pipeline produces significant numbers as indicated in Table 4.2, the IT industry has absorbed little of the output. In 2004 there were only about 70,000 workers in the IT industry as a whole, with only an estimated 16,000 employees in the software export sector.[22] The educational system has not been re-engineering itself as quickly as the Indian and Chinese competitors. For example, as of 2002 no higher-education institution used the complete reference models of ACM or IEEE for computer science or Management Information Systems.[23]

Russian software exports are the smallest of the Big Three at roughly 350 million USD in 2004. Of this amount, about 50 million USD came from captive software R&D (i.e. owned by foreign firms) and roughly the same amount in software products. Indigenous software product firms have not been influential, with the exception of Kaspersky Labs, a global provider of anti-virus software.

Russia has a respectable presence of American R&D centers including Intel, Motorola, and Sun, as well as some European firms, such as Nokia. Other foreign firms perform contract R&D. But, in total, this presence is significantly smaller than multinational corporations' presence in India and China.

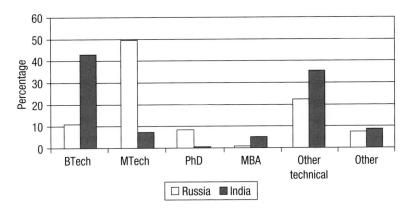

Figure 4.3 Educational background of employees in Russian and Indian firms.
Source: Bardhan and Kroll.[24]

Table 4.2 Russian educational system's annual supply of IT labor

	2002–2003	2003–2004
IT engineering graduates	42,000	46,000
Math and Physics majors	22,000	22,000
Non-IT engineering graduates capable of entering IT workforce	69,000	76,000
Graduates from other disciplines capable of entering IT workforce	71,000	81,000
Total fresh IT labor supply	203,000	226,000

Columns do not total due to rounding.
Source: Auriga.[25]

Nevertheless, the Russian software export industry has been growing steadily with the consequent growth pains familiar in China and India: tight labor markets for the best talent, increasing wages in the major cities, and high rents in desirable locations. Russian firms have been quick to emulate the Indian model by embracing international quality standards such as ISO and CMM. Motorola Russia attained CMM Level 5 in 2001. One of Russia's largest independent IT services firms, Luxoft, attained CMM Level 5 in 2003.

The internal domestic marketplace has become healthier due to the economic boom of the early 2000s, and more sophisticated. This is mixed blessing for offshoring. As is typical at this stage of industry growth in emerging nations, domestic customers are more profitable than foreign customers, diverting companies' attention to domestic markets.

One of the most important limitations to industry growth is the relatively small size of Russian firms. Only a handful of firms are larger than 200 employees. Individual firms cannot – or do not want to – grow for a number of reasons: lack of capital (capital markets are immature) and a desire to stay off the radar screen of the erratic and punitive Russian tax collectors. There is an equally strong desire to stay off the radar screen of the endemic organized crime and burglary rings. Finally, there is a Russian cultural propensity to work in intimate family-like organizations. The result is that software firms choose to operate in a constellation of alliances, sub-contractors, and consortia that allow them the ability to accept new work without growing their core workforce.

The other two "I"s

The three software stars of the 1990s were the "three 'I's." India was covered earlier in this chapter. We now provide brief sketches of the other two "I"s: Israel and Ireland, with the monikers of *Silicon Wadi* and the *Celtic Tiger*, respectively.

The two countries have several characteristics in common. Both nations are small with strong educational systems. Both became successful, in part, in software products, which is unusual for offshore nations. Both have seen their starring role in software dull somewhat after the technology boom ended. Both attracted significant investment from major global technology firms (although into different segments as we discuss). Both have relatively high wages that make them less attractive to offshoring of IT services with its emphasis on wage differentials. Both have been hurt by the fierce global competition as companies look to India and China. Both countries did not grow the software "contract R&D" segment that India has been so successful at, preferring to conduct software R&D on their own, or within foreign subsidiaries.

Israel

During the technology boom years of the late 1990s, Israel's software strengths were in synch with global interest. With a highly educated labor pool and tuned-in technology entrepreneurs, the country was able to grow many innovative software product firms, of which the most well known is Check Point, the information security firm. Israel became the number one foreign destination for American venture capital and the home of more than 100 technology firms listed on the US NASDAQ stock market.

Until the early 2000s Israel was the preferred destination for offshore software R&D, with nearly every major American technology firm having some research presence in country. The size of these R&D centers is substantial as measured in number of employees:[26] Microsoft, 400; Cisco, 500; Intel, 2000; HP, 2000; and IBM, 500. And all these centers are performing advanced R&D work, rather than the lower value activities that are often transferred offshore. For example, the Intel Centrino processor was designed at Intel-Haifa. In addition, small software firms, with easy access to capital, continue to emerge and be acquired by foreign giants, such as the 2001 HP acquisition of Indigo for 900 million USD, or the 2004 acquisition of tiny Actona Technologies by Cisco for 80 million USD.

Ireland

In the 1990s, Ireland attracted investment from 120 foreign technology firms that came to take advantage of its English-speaking population, low wages, tax holidays, its location as an entry point to Europe, and the ease of doing business there. Half of the foreign technology firms were American, including Microsoft, Oracle, IBM/Lotus, Symantec, and Sun. Ireland performed some offshoring services, similar to India, for companies such as EDS, Xerox, and IBM. And, most notably, Ireland developed a healthy sector of home-grown innovative software product firms that was the source of its pride: Trintech (e-payments), Iona (middleware), Riverdeep (educational software), and Baltimore (encryption for digital security). By 2004 Ireland's software industry had 30,000 workers in more than 800 international and indigenous software companies.

One of the key differences between Israel and Ireland is that while many of the foreign firms in Israel were conducting advanced software R&D, the foreign technology companies that came to Ireland did not transfer high-end knowledge, and were performing low-end activities of localization, porting, assembly/packaging, and logistics distribution. Once Ireland's wage and cost advantages disappeared, many of these multinationals departed for lower-cost nations. By 2003 the head of Ireland's software industry association urged the sector to focus on niche areas of software in order to remain competitive on the global stage. Within the software product sector only the e-learning niche seemed to represent a strong, viable cluster of firms. Most of the other home-grown firms were not been able to grow to become globally competitive, with only 24 of 700 Irish software firms reaching annual revenues above 2 million euros.

Six lesser knowns

In addition to the more successful software nations, there are many more countries offering offshore services. They differ considerably: some are very large, and others are tiny. Some are located nearby, others very far away. What they do have in common is that they are relatively unknown. Even though their software export sector is often small in size, these countries should not be ignored. On the contrary, a closer look at these and at others may prove fruitful because each of them offers a specific set of advantages to potential users.

Latvia

Latvia, one of the three Baltic states, was one of the 'Silicon Valley' clusters within the former Soviet Union.[27] Its software specialists were well regarded. During the Soviet era, Western software licenses could not be purchased, due to the stringent CoCom regulations. Instead, software was bought or pirated via third countries and then distributed throughout the Soviet Union. For example, there were Latvian centers for the adaptation and distribution of products from the German firms Siemens and Software AG. This paid off as soon as the Soviet bloc began to splinter. Through contacts with Latvian emigrants working in Germany, the first software orders came from German customers. Clients included Siemens Nixdorf and the German State social insurance office, which used Software AG products.

After Latvia regained independence, nearly all state scientific institutes were disposed of and the big industrial enterprises fell to pieces. The work with German organizations helped to retain many of the newly unemployed software professionals. Today, the focus continues to be on Germany, for which Latvia is a nearshore destination (it is only a 2-hour flight). Latvia also has an advantage in widespread knowledge of the German language, resulting from 700 years of German economic influence and, from time to time, German political dominance in the country. The largest offshore provider in Latvia is DATI, which is also one of the biggest software houses in Eastern Europe. It works for clients such as Software AG, insurance company AXA, and the largest German telecommunications firm Deutsche Telekom.

Around 50 Latvian IT companies are actively engaged in offshoring, exports were 20 million USD in 2001. Latvia's 2004 membership in the European Union (EU) will make business with customers in Western Europe far easier. Far more liberal work permits will allow easier mobility.

The Baltic IS Cluster was formed in 2001, in order to expand exports from the three Baltic states, along with nearby Belarus. The Latvian information technology and telecommunication association (LITTA) is the coordinator of this regional cluster, which has 20 members.

Romania

In the 1930s Bucharest was known as the Paris of the Balkans. Although the city has lost that luster, it is a vibrant hub of many software firms. Within the *transition economies* of Eastern Europe, Romania has capitalized on its low wages, comparable to Indian wages. It is the largest exporter of software in the former Soviet bloc besides Russia. EITO estimated Romanian exports in 2003 at 130 million USD. The country has 370 firms exporting software, of which most are exclusively exporting.[28]

The linguistic roots of Romanian as a Latin language means that English – and especially French – are spoken more widely than in other Eastern European nations. Language has also allowed its firms to move into the IT-enabled services segment. A case in point is Softwin, one of the larger IT firms, with 400 employees. It has built practices in four related areas: software services for exports, software services for the growing domestic market, product software and IT-enabled services, including contact centers for French firms, and domestic e-publishing.

A number of foreign technology firms built R&D centers in Romania such as European giants Alcatel and Siemens, each of which has about 400 employees in their Romanian centers. Alcatel is said to export 10 million USD per year in software from Romania.

The industry growth, although consistently in double digits, is stunted by an economy with poor infrastructure, lack of capital (and little venture capital), and a continuing brain drain of the best graduates of its technical and engineering schools. EU membership expected sometime before 2010 will likely address some of these issues. However, EU membership will also lead to a rapid rise in software wages, which will erase one of Romania's main competitive advantages.

Malta

Malta, consisting of a few islands in the centre of the Mediterranean, is with 400,000 inhabitants one of the smallest countries in the EU. Located between Sicily and Tunisia, it is an important tourist destination. Malta is a typical Mediterranean country but with a difference: because of its long historical association with Britain, there is a very good command of the English language. Many Maltese also speak Italian, French, or German. Although it is in Southern Europe, the business culture and work ethic are more 'northern' and somewhat resemble the British.

Although most Europeans are surprised to hear that Malta is a nearshore IT location, the country offers some specific advantages. The time zone is the same as Amsterdam, Rome, or Frankfurt. Malta is pleasant to visit and because of its small size, it is easy to meet all major offshore providers. A group of Belgian managers, looking for an outsourcing partner in 2004, were able to meet all their potential candidates in just 1 day. Malta joined the EU in 2004 and the political, economic, security, and legal risks are very low.[29] It has a modern infrastructure and programmer wages are much lower than in Western Europe.

The country now has a few dozen software firms, both local as well as foreign-owned. Due to historic ties, several British companies offshore to Malta. An example is Safeway Stores, a large food retailer, which operates more than 500 stores in the UK. Since 1998, it has outsourced work to Crimsonwing on Malta, which is connected to Safeway's mainframe. Over the years, a 30-person team has delivered more than 200 projects in areas of merchandising, supply chain, and corporate systems. Crimsonwing is one of the largest software houses on Malta and employs 130 IT staff. It has several British customers, such as Barclays Bank and Securicor.

Vietnam

With a population over 80 million, Vietnam is considered a "young country" as 60% of the population is younger than 25 years.[30] Unlike many other developing countries, the literacy rate is high, at 94%. Vietnam's strength is in its inexpensive labor costs, the average wage rates are one-third to a half of the costs of similar skills in India, and with lower-turnover rates. There are about 100 organizations, including universities, colleges, and institutes that are training IT students. These organizations are producing about 3500–4000 IT students every year, which is still a relatively small number. IT training centers are now being set up in the large cities, including some sponsored by the large Indian providers TCS, NIIT, and Aptech. TCS has even deployed Vietnamese programmers at one of its centers in Mumbai.[31]

Socialist Vietnam is one of the newer entrants in the global software business. The results of its *Doi Moi* reform movement of 1986 have yielded economic improvements in many fields, including informatics. Improvements have been made in the field of Internet connectivity, although the prices for bandwidth services are still high. Vietnam has 10 software parks spread around the country in its major cities: Ho Chi Minh City, Hanoi, Danang, Haiphong, and Hue. These software parks offer tax holidays and other incentives to software companies. The most successful one is the Quang Trung Software Park in Ho Chi Minh City, which houses more than 50 companies employing more than 1500 software engineers.

The software industry has grown rapidly – at a 30% rate in 2003. Vietnam has more than 400 software companies, employing about 10,000 software engineers. However, almost all of these enterprises are small, and generally still weak in terms of project experience, business knowledge, management, and English. The Ministry of Post and Telematics, together with Vinasa (the Vietnam Software Association), have taken steps to increase quality standards for the industry. There are about 15 companies with ISO-9000 certification. And one organization, FPT's software division, the largest IT enterprise in the country, is certified at the CMM Level 5, the highest level of this international quality mark. The government plays an important role in other respects as well: it actively promotes the use and development of Open Source Software.

Foreign clients include Japanese, European, American, and Indian companies, including Cisco, Nortel, IBM, Sony, and Bayer. The US is the major client; the existence of the large "Viet Kieu" diaspora has been helpful to bridge the gap between Vietnam and the US. UK-based Harvey Nash, an IT services provider, is active in Vietnam. Japanese corporations have been targeted as an important new market, and Vinasa predicts that the revenue from software exports to Japan will reach 5 million USD in 2004. A Japanese language center is planned to help programmers work more easily with their Japanese counterparts.

Vietnam's software exports were estimated at 30 million USD in 2003.[32] The government set an ambitious, and perhaps impossible, goal of increasing the export of Information and Communications Technology to 300 million USD by 2005. It also plans to establish a trade promotion organization to help software producers find foreign customers.

Bangladesh

With a population of 130 million people, Bangladesh is one of the largest developing countries in the world. Similar to Vietnam, Bangladesh offers very low wages and a relatively large labor pool from which to draw. Due to its colonial British heritage, English is widely spoken among the educated classes. More than 40 public and private universities, and some institutes and colleges, are offering degree courses in the area of IT. Every year, 3000 IT graduates are coming out of these institutions. In addition, there are large numbers of IT training centers with an estimated total output of

12,000 per year.[33] Students from Bangladesh University of Engineering and Technology (BUET), Dhaka University, and some private universities have scored high marks in international computer programming contests. Basic technical knowledge is considered to be adequate. Several small foreign clients reported that they could easily attract qualified programmers, which would have been more difficult for them in other countries.

There are more than 200 software firms in Bangladesh, although selling hardware is often a significant part of their services. Most of these IT companies are small: a company of 50–100 people is considered large. Several dozen Bangladeshi software firms are doing work for foreign clients, but there are very few subsidiaries of foreign technology companies. Bangladeshi software exports were estimated at 5 million USD in 2004.[34]

Corruption, which is a major problem in Bangladesh, has hardly any impact on the international projects. We know of a Dutch company that paid a small amount in order to speed up the process of acquiring telephone lines for its subsidiary. This was quite reasonable, given the fact that it can take local citizens up to 27 years to get a telephone connection.

Several Dutch organizations have outsourced software work to Bangladesh. Compared to The Netherlands, cost savings of more than 50% (including overhead) can be achieved. These clients mentioned that the quality of documentation and testing is sometimes inappropriate and quality assurance is weak. This results in software which needs to be sent back for debugging and reprogramming. There is always a risk that the engineers leave for a position abroad, but they can be convinced to stay if the project is challenging (e.g. using new technology) or if better working conditions, additional training, or other incentives are offered.

Unlike Malta or Costa Rica, Bangladesh is not a tourist destination. Foreigners who stay in Bangladesh for a longer period of time consider the country, and especially its capital Dhaka, messy, noisy, and unattractive. When a Dutch company needed a project leader to head its team in Dhaka, it selected a person who used to travel to such countries.

Costa Rica

Costa Rica is one of the most stable and long-standing democratic countries in the world. It has been a democracy without interruptions since 1889, which is very unusual for Latin America. It is one of the few nations without an army, and public education has been a priority in government spending. With a population of only 4 million inhabitants, it has a pool of four public universities and many private universities. It has the best education system in Central America, along with the highest literacy rate (95%).

Since 1997, investments of Intel Corporation have turned Costa Rica into an important exporter of computer chips. In 2002, Intel also opened a software division in the country. There are around 100 software companies, most of which are small or medium enterprises. Software exports reached a total of 70 million USD in 2002.[35]

The majority of these exports (63%) are destined for Central America, where regional competition is limited. Some firms are conducting nearshore software services for North American companies.

In order to diversify exports, the government has targeted IT. An alliance of IT-related organizations started the program, "Costa Rica: Green and Smart ..." This slogan is derived from the unique natural beauty of the country, which houses 5% of world's known biodiversity. The program includes activities to promote the development of human capital and to improve the visibility of the country abroad. Two of its interesting thrusts are exploiting synergies with the large Latino market in North America, and development of IT-enabled services.

Part II

Managerial competency

5 Offshore strategy

Erran Carmel and Peter Schumacher*

Offshoring has become the management fad of the moment; the *innovation du jour*; the accepted reaction to cost pressures; or before 2000, during the technology boom, the reaction to labor shortages in industrialized nations.

Does any of this constitute a *strategy*?

In order to begin answering this question, the notion of "offshore strategy" needs to be qualified for each company. It needs to be seen in a broader context and not in isolation. First, for medium- and large-sized companies, offshoring is part of a broader globalization strategy. Second, also for medium- and large-sized companies, offshoring IT is just one part of the larger scope of *knowledge and service activities* that companies source globally, along with IT-enabled services (ITES) and research and development (R&D) of all kinds. Thus, strategic offshoring is, but, part of an overall sourcing strategy which deals with the portfolio of strategic options: whether these options be at home, abroad (in both high-wage and offshore nations), inside the company, outside the company (outsourcing), and through various collaborations.

Accordingly, an exacting definition of *IT offshoring strategy*:

> The proactive logic, evident to an outsider, in a firm's portfolio of IT offshoring activities within the firm's larger scope of global sourcing.[1]

Admittedly, few companies take such a holistic view of their offshoring strategy. Most companies offshore with a more isolated set of goals, and with a shorter time horizon, in an evolving, or even reactive strategy. US-based GE is an exception in that it did articulate a broad strategic vision for offshoring. The case of GE is described in detail at the end of this chapter.

This chapter focuses on corporate strategies that are enhanced by, or that are unique to, offshoring. This is an important distinction because offshoring is often confused with outsourcing. Companies can also outsource to a provider at home, in their own country. To rephrase: this chapter introduces a topic that has not received enough attention: *What is different about an offshoring strategy?* Separately, this chapter does not examine strategy from the perspective of the offshore provider (which provides services, such as an Indian provider or a European firm with in-house offshore resources; more on the providers appear in Chapter 1 and later in Chapter 11).

* Schumacher is at Value Leadership Group Inc., Germany.

The chapter begins by looking at the cost-reduction strategy. We then move beyond cost-reduction and introduce the notion of strategically leveraging offshore. This is followed by an examination of the strategic perils of offshoring. Finally, various strategic collaboration strategies are introduced.

Cost-reduction strategy

Since offshoring has been driven primarily by cost savings, does this constitute a *strategy*? Is offshoring a strategy if the company squeezes a bit more out each euro that is allocated to IT wages? Strategy theorists will disagree and debate the nuances of such definitions. Unlike a new product line, or new improved services, offshoring does not constitute a company's *competitive strategy*, since its goal is merely to increase operational efficiency.

IT offshoring has been driven primarily by the executives' desire to lower operational costs. Lowering operational costs does not necessarily translate into a company's strategic advantage, just as saving money on a new office lease is not a strategic advantage, but merely the relentless day-to-day effort of any company to reduce its operating costs.

However, in some industries IT offshoring is beginning to be viewed as a *strategic necessity*. We have heard this expressed in stronger terms: "offshore or die." When one company's cost efficiencies allow it to lower prices or expand its competitive options, then other companies must match their competitor's strategy, or fail. Offshoring is becoming part of the larger environment of *hyper-competition*: companies are swept into faster and faster cycles of competitive responses and reactions in order to remain financially viable and cost competitive. *Not* offshoring may well become a strategic peril. Such was the case of one of America's largest television manufacturers, Zenith Electronics, which resisted offshoring for decades, while slowly shrinking, before it disappeared completely.

The *de facto* entry point for the offshore strategy is cost reduction. This was introduced in the Offshore Stage Model, in Chapter 1, and shown again here in Figure 5.1. Most firms that have offshored will likely remain at this stage (stage 3) since their only strategic goal is cost reduction.

We have heard some executives pursuing a cost-reduction strategy ask "How much should I offshore?" By this they mean: What is the ultimate ratio of resources that should be offshore (versus onshore)? In practice, this translates into one of the following two measures:

- The ratio of IT headcount (IT staff) that should be offshored.
- The ratio of IT budget that should be offshored.

Among major American companies the trend-setter in this style of numeric goal setting has probably been GE, with its successful offshoring strategy. Jack Welch, the revered former CEO, is said to have established the 70:70:70 rule; 70% of all GE's IT work should be outsourced, of which 70% should go to global preferred vendors, and

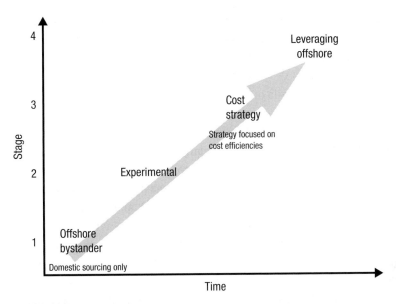

Figure 5.1 The Offshore Stage Model (repeated from Chapter 1).

of that 70% should be done on the vendor's premises (which may be offshore). Over the years, the influential 70:70:70 rule was embellished by the offshore industry and it is usually retold like this: 70% of all IT work should be outsourced, of which 70% is to be outsourced offshore, and of that 70% should be in India.

Other firms have followed GE's lead. We learned of similar numeric objectives at three large organizations. In all three cases the numeric goal was, roughly, to double offshore staff within just a few years:

- A major American financial services company expected to reach 18% of headcount offshore by the end of that year. The goal of 40%, to be reached soon after that, was suggested by a consulting firm.
- A large American embedded software firm had 9% of headcount offshore, with a goal of 15–20% offshore within 3 years.
- STM, the large French–Italian chip maker performs applied R&D offshore (though its main fabs, its manufacturing capacity, remain in Europe and America). In 2002, it had 24% of its engineers in low-cost Asian locations (mostly India) and 13% in low-cost European nations, such as Russia. The target for 2006 was to increase the relative percent in low-cost nations to 40% and 24%, respectively.

Such numeric targets are a rational means for implementing the cost-reduction strategy because once that strategy has been articulated, it needs to be operationalized by setting goals, defining objectives, and linking them to rewards and performance reviews. Not all firms have such numeric goals; but without them, organizational resistance may be more difficult to overcome.

There is no magic number for the "How much should I offshore?" question. Numeric offshoring goal setting should be less important than the company's overall global sourcing strategy and its relative success in implementing its global objectives. Smart companies need to see *measurable* financial results or other strategic benefits and adjust the "How much" incrementally.

Leveraging offshore strategically: beyond cost savings

Leveraging offshore strategically applies to those companies in the fourth stage of the offshore stage model (Figure 5.1). While relatively few companies have advanced into this stage, such companies gain more from their offshore strategy than using offshore locations simply as low-cost suppliers.

In particular, the spotlight in this section is on those strategic goals that are enhanced or unique to offshoring. While there are many types of strategic advantages to outsourcing, to acquisitions, and to collaborations – all of which can also be performed in your home country – this section separates what makes offshoring truly different.

Strategic offshoring may be even more important for companies that perform software R&D than for "end-user" organizations like GE or Deutsche Bank. Hundreds of Western high-tech companies, European, American, and Japanese, are doing software work offshore in either captive centers (subsidiaries) or via outsourcing, which in this case is also labeled as "contract R&D." For example, Indian-based Wipro provides offshore contract R&D on a massive scale. It has 6500 engineers who supply services to many technology firms including 9 of the 10 largest global telecommunication operators.[2]

Six strategic goals to leverage offshore are introduced here (and depicted in Figure 5.2). Not all of these apply to both types of software work: end-user companies (such as banks) and software R&D (product or embedded). Of the six strategic goals, one applies only to end-user IT, two apply only to software R&D, while three apply to both types of work.

1 Speed, agility, and flexibility

The first strategic goal is illustrated with the case of the German airline Lufthansa:

> *In 1999, Lufthansa Cargo embarked on a massive IT project: to automate its booking system. The company was hesitant to perform this project in-house because it was not certain it had sufficient resources and sufficient know-how. The project required expertise in middleware with specific expertise in the middleware software product BEA. On top of this, the air carrier, under competitive pressure, wanted the project done very fast.*

> *Lufthansa put the project out for bid. Two of the largest US-based IT service firms each bid about 50 million USD, committing to a 2-year development duration. The third bidder, Perot Systems, like the other two bidders, brought expertise in airline systems to the table. But Perot relied on offshore resources to staff some of the project. It won the bid for just 25 million USD with promised*

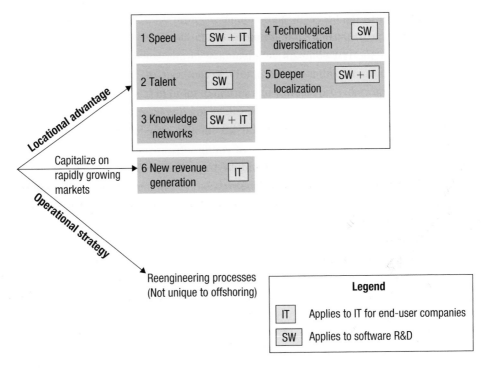

Figure 5.2 Leveraging offshore: strategic goals of offshoring.

delivery in only 9 months, which is less than half the duration of the other bidders. Perot bid without some of these resources in-house, knowing that if they win the contract they can "buy" the resources in the Bangalore cluster. They did.

Lufthansa Cargo became the first air cargo carrier in the world to offer its customers an online-booking system.

The Lufthansa Cargo case illustrates the power of offshoring: the ability to draw from a large labor supply to achieve quick ramp-up time (the time get the project started) and reduced project duration (time-to-completion).

Certainly, speed is achievable in the US and Europe, but it is quite expensive. Moreover, in Europe, quick ramp-up for large projects is more difficult because labor is inflexible and largely immobile. Gathering engineers together from distant European locations is unusual. In contrast, one of the attractions of offshoring is the ability to build massive centers of hundreds, and sometimes thousands, of software professionals. In other words, offshoring is an opportunity for considerable consolidation in order to achieve economies of scale.

Reducing time-to-market can also be achieved using follow-the-sun development (introduced in Chapter 1). By taking advantage of time zone differences, the offshore unit can accelerate a project: while the British workers sleep, the Asian offshore unit is refining the prototype and then passing it on for inspection, feedback, and refinement at the end of their day. If the company in question is a software product company, it may be able to shave several months from the development cycle and release a product earlier, giving it a competitive advantage. Some software companies have benefited from the speed advantages of follow-the-sun. For example, Portal Player, a maker of multimedia chips and embedded software for Apple's iPod, with R&D in India and Silicon Valley, was able to perform rapid prototyping using follow-the-sun.

Offshoring organizations can be speedier and more agile due to the large, motivated supply of labor. This is a labor force of young software engineers that are driven very hard to succeed. They work long hours, often sleeping at the office to get more done. Thus, the organizations offshoring to these destinations can afford to assign a large number of engineers to a problem; and these engineers can be assigned to forge ahead in several directions instead of just one; they can ramp up and respond to a business need within days instead of months. Infosys staffing illustrates the depth of the labor pool. Infosys, as one of the largest Indian providers, receives 900,000 job applications per year. When the company needs to staff more projects, it turns opens the labor pipeline a bit more. Murthy, CEO of Infosys, asserts that his company is "always selling itself at twice its current size."

2 Talent for innovation

Talent has a specific meaning in the software business. Of the millions of software professionals in the world, most are "blue-collar" programmers with commodity skills. They are largely interchangeable with one another. A blue-collar programmer in Bulgaria or the US or Brazil is undifferentiated (except for his price).

However, the top of the software labor pool is the *talent*. In software, as in music, anyone can learn to play an instrument, but only the talented can compose a symphony. The software talent are those who can innovate. They are the brilliant programmers and software architects. They are not interchangeable. Nor are they easy to find. For software R&D this talent is a key competitive factor.

In previous decades, software companies tapped talent locally, in their own nations. In the 1990s, they began turning to Israel, India, and later still, to China. By the early 2000s, there were several hundred software R&D organizations in the leading offshore nations.[3] Intel, for example, has R&D organizations in India, China, Russia, Israel and several other offshore destinations. While some of the offshore tasks are routine, and some of the benefits are in lower wages, Intel located and expanded in these locations in order to access the local talent. Some of Intel's leading chip sets are now designed outside the US, by engineers in Israel and, more recently, in India. A measure of innovation is the generation of patents: engineers within American technology firms operating in India filed a cumulative one thousand (US) patents by the end of 2003.[4]

Software companies prefer to set up centers in locations that are a magnet for talent, in technology clusters. And this becomes a virtuous circle: companies come because of the talent; the talent comes because the best companies are there. A technology cluster is an ecosystem that attracts key technology ingredients: the best technical and managerial talent, as well as other ingredients, such as capital. In the 1990s, the undisputed global technology cluster was Silicon Valley. Although it was expensive, the Valley attracted technology firms which were interested in access to its talent and access to critical local (soft) knowledge that it had to offer. The up-and-coming technology clusters of the 2000s are in India and China.

3 Building global networks for knowledge sharing

Successful companies encourage their various local and offshore centers to connect and share in all kinds of ways. These locations collaborate, share knowledge, offer ideas to each other, learn from practices in other countries, and solicit small problem-solving solutions from each other. The importance of these networks applies to all kinds of global companies: to non-technology multinationals, such as Unilever; to pharmaceutical companies, such as Eli Lilly; to technology companies, such as Intel; and to software companies, such as Microsoft.

4 Technological diversification

Large technology firms can diversify their technology portfolio more effectively when they spread their R&D facilities globally. Granstrand and colleagues[5] studied global technology firms and found that the companies that attain long-run competitive advantage are those that have expanded to many foreign locations and, in the process, achieved technological diversification. These corporations are leveraging their offshore units to attain strategic diversification.

Companies need to have a diversified set of technological competencies not only in the firms' distinctive *core* competencies – those that outsiders are likely to recognize – but in three others. The first of these competencies is in "niche" areas, which are intrinsically small, and are those in which the firm has less expertise, a lower profile, and fewer resources. The second competency is in background areas, dealing with processes and coordination, allowing firms to benefit from technical change. For example, the emergence of India as a global center for applying mature quality processes in software development is a strategic "background competency" for some companies. Third, companies may also retain competencies in some marginal areas in which they have no distinct advantages, although these generally tend to be outsourced.

5 Deeper localization

Almost all software has to be localized to local language and culture. The closer you are to your customer, the deeper the localization. In strategic parlance this is called *local responsiveness*. Many companies localize by hiring foreign language experts at home. However, situating localization in the target market allows firms to better customize products to the local markets, particularly the large and more promising markets, such as India and China. This is the case with Microsoft's significant presence in China, with its development centers in Shanghai and Beijing which devote significant resources to Chinese language scripts and other local needs.

6 New revenue generation

Since the offshore markets are growing much faster than those in industrialized nations, a company that is offshoring its IT functions has greater opportunities to generate new revenue and new value from these operations. The path that many companies have taken is to spin off their offshore center, create an independent offshore unit that can sell IT services or products to third parties. A number of firms have gone beyond that point. Offshoring has given creative players greater opportunities to capture value from their operations, particularly in India, by selling assets at a multiple of their original investment. One offshore expert summarized it this way: "You take a look at your assets, polish them in India where it is cheap, and sell them dear."

Three examples illustrate this strategy:

- British Airways created its Indian-based WNS division in 1996 to reduce its operational costs in IT and IT-enabled services. WNS was quite successful and grew dramatically. British Airways recognized the strategic potential and began selling off pieces of WNS. Its software assets were sold to Kale Consulting. What remained of WNS was a successful IT-enabled services firm. In 2002, a 70% stake was sold to Warburg Pincus. Today WNS is the largest independent IT-enabled services firm in India.
- US-based Citibank created its Indian-based CITIL division in 1989 with an initial investment of about half a million USD. The division provided IT services to the parent and, in parallel, began developing banking software products. In 2000, Citibank renamed the division i-flex to capitalize on the brand recognition of its Flexcube software product. i-flex became a public company in 2002 and later reached an incredible market capitalization of 1 billion USD.
- UK-based ebookers, a large travel company, created its Indian-based Technovate division in 2001 to support the parent in IT and IT-enabled services. By 2003, ebookers recognized the potential strategic value in its successful division and began selling Technovate piece by piece, with the first piece going for 10 million USD based on an impressive market valuation of 160 million USD.

Operational strategy

Offshoring may also be used as a strategic opportunity to attain important operational goals, such as re-engineering internal company processes. Corporations have traditionally used the occasion of building a new information system as an opportunity to redesign wasteful, inefficient corporate processes, such as account processing and customer approvals. Over the years these opportunities for organizational transformation also coincided with outsourcing.

Does offshoring offer a unique advantage in this regard? No, because there are no strategic locational advantages to the offshore geographic locations, the offshore providers, nor the offshore labor. In particular, the offshore providers have no unique advantages in organizational transformation. While the offshore firms have been effective at delivering generic services, skills, and bright programmers, these firms have no advantages at re-engineering processes. Offshore providers have not developed advantages in various vertical fields and industries relative to the American or European firms with which they compete. We should not confuse the high-quality processes practiced by Indian providers (e.g. Capability Maturity Model (CMM) Level 5) with the ability to innovate the client's performance or system capabilities.

While there are many strategic advantages to outsourcing, we do not see evidence that companies can attain better operational strategies offshore simply because it is offshore. We illustrate this point by examining the related area of outsourcing strategy (not to be confused with offshoring strategy). The *Outsourcing Journal* gives out annual awards to IT outsourcers that achieve such goals as "most transformational" and "best process improvement." These are strategic benefits of outsourcing which can be implemented by American, European, Indian, and other firms. However, there is no inherent advantage in offshoring to achieve transformational change unless that goal is combined with other unique offshore advantages such as speed.

In conclusion, there are six strategic goals in leveraging offshoring that go beyond cost reduction. The common denominator in five of the six goals is the ability to take advantage of *location-specific* factors. These locational advantages are in human resources, easier links to various geographic locations, and proximity to markets. These factors are called location-specific because, by and large, they cannot be moved. The offshoring company leverages these location-specific factors using its own know-how in order to create strategic value.

Strategic perils

Companies need to be cautious of strategic missteps in offshoring: losing their core competencies, forgetting their strategic goals, and losing advantages in proprietary knowledge and proprietary code. We cover each of these perils in this section.

Core competency is lost

> *"The distinction of core versus non-core activities does not hold in our case."*
>
> Senior IT manager at a large
> North American retailer on the firm's offshoring plans

> *"We have very few competitive technologies in our IT."*
>
> Senior manager at a major Wall Street firm
> regarding the firm's offshoring plans

The firm's *core competencies* are those capabilities that are the source of its competitive advantage over its rivals. These are capabilities that its competitors cannot imitate – at least not without great effort. Today, the conventional wisdom is that companies should hold tightly to their core competencies while trimming the rest – becoming "virtual" or "hollow" corporations via outsourcing, or more recently, via offshore outsourcing.

It is clear that smart companies should never outsource their core competencies, for they will lose them. But what in IT is really a core competency? This is a question that companies have been struggling with for years. Information systems that only a few years ago were considered "strategic systems," and therefore verboten to outsource, are no longer considered vital as the quotes above illustrate. Companies are much more liberal in outsourcing a variety of activities. Without doubt, some companies see IT as a corporate function that has no core competencies ("IT doesn't matter").

The enthusiastic headlong plunge of some companies into offshoring may cause decision-makers to overlook their core competencies in IT. We note Strassman's studies of outsourcing in this regard. In 1995, Strassman headed a study that led to the writing of an influential article called "Outsourcing is for losers." He followed up on that article, releasing a study in early 2004 titled "Most outsourcing is still for losers."[6] Strassman calculated the value added of outsourcing using financial statements of two sets of data: US firms in general, and of US banks (who tend to be aggressive in their outsourcing strategies). In both cases he found that the firms that outsourced performed less well. Strassman argues that his results suggest that "companies already failing for other reasons tend to outsource increasing amounts of work, thus diminishing their value added."

Some companies have examined offshoring vis-à-vis core competencies and determined that some competencies should not be offshored, drawing a line around a set of technical and business competencies. For example, a major American health care systems firm articulated the capabilities that it would *not* offshore: domain knowledge, architecture, integration, and delivery. Companies can also approach this analysis by conducting an inventory of their corporate-wide systems and determining which systems embed core competencies that the firm should not outsource onshore or offshore.

Forgetting the broader strategic goals

Just as core competency is ignored, the plunge into offshoring may ignore a company's broader corporate strategic goals. We present two non-IT examples from the consumer goods industry that illustrate how offshore cost reductions need to be balanced with other strategic goals.

Many firms in the athletic shoe industry outsource much, if not all, of their manufacturing to low-wage nations (Nike, Reebok) in order to reduce costs. New Balance has taken a different approach.[7] New Balance maintains about 20% of its production in the USA. Its costs in the US are somewhat higher, though its US productivity is also higher. Producing domestically gives it the advantage of closer integration between design and production, and greater quality control. The higher costs represent only 4% the costs of a typical shoe.

Many firms in the apparel industry offshore to reduce production costs; yet Spanish fashion chain, Zara, determined that time-to-market is more important than reduced labor costs and kept most production at home in Spain. In the fickle world of fashion taste, being flexible and responsive is essential. By coordinating the production and logistics process very tightly through its IT, Zara is able to restock its European stores twice a week with completely new styles, with time-to-market that is 12 times faster than its rivals.[8]

The lesson in these stories is that a focus solely on offshore cost reduction may divert attention from the larger, more important strategic goals.

Losing advantages in proprietary knowledge and proprietary code

Software companies, in products and embedded code, face significant offshore risks involving code theft or leakage of proprietary knowledge (as covered in the risk section of Chapter 2). Stolen code that ends up in the hands of a competitor may be a severe risk. Knowledge leakage is insidious. If a firm is offshoring to a growing competitor nation – and many of the Asian destinations are in this category – then it runs the risk of transferring key know-how abroad, to its eventual rivals.

American technology firms have staked out positions at all ends of this strategic spectrum. "Competence will stay in US," remarked one R&D manager in a conversation with us, meaning that at this firm there will be tight control over what is offshored. In 2004, Microsoft was quick to deny that core parts of its next Windows software release will be offshored to India. However, in the same year a study found that 79% of American software companies that were offshoring were performing some *core* development tasks offshore.[9] Often, it is the American technology giants such as Intel and Motorola that have been moving core development tasks to China and India.

Strategic collaboration: offshore business models

Once the offshore vision is articulated, the offshore strategy needs to be developed, defined, and executed. This execution and implementation can travel down one of two paths: "Buy versus Build."

Certainly, "Buy versus Build" is not unique to offshoring: it is a generic business decision that crosses all industries and all business types. Offshoring is no different: variations on the offshore collaboration strategy diagram of Figure 5.3 have been used in countless corporate meeting rooms and consultants' presentations.

"Buy versus Build" represents a set of trade-offs. The "Buy" strategy encompasses offshore outsourcing or offshore out-tasking (project contracting). From the point at which the decision is made, "Buy" implies faster ramp-up time because the provider already has operations in place. It is a less risky short-term strategy, particularly because of the difficulties of international business; and it generates greater short-term savings. Generally, the "Buy" strategy is chosen by firms that are offshoring secondary corporate functions, such as IT or IT-enabled services.

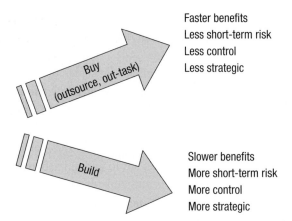

Faster benefits
Less short-term risk
Less control
Less strategic

Slower benefits
More short-term risk
More control
More strategic

Figure 5.3 Offshore collaboration strategies.

The "Build" strategy (really a build/own strategy) represents a larger investment upfront and longer time-to-benefits. It is riskier in the short term, but the firm enjoys greater control which is paramount if core competencies are to be offshored, or if intellectual property leakage has serious consequences. The "Build" strategy is superior in managing knowledge transfer, when creating the proper offshore organizational culture is deemed important. Finally, if firms are leveraging offshore strategically, then they can attain new revenue sources from the "Build" option, by spinning off their units or spinning-off products.

The disadvantages of "Build" are in longer lead times, bureaucratic and corruption-related hassles to set up and run offshore operations, and the need to recruit and hire. Small firms are likely to be hurt by attrition as their engineers are lured to larger firms that are able to offer more prestigious work and, often, higher wages. Firms have two "Build" options: acquire an existing firm offshore and then grow it, or "greenfield" – build a subsidiary from the ground-up, sometimes by securing the land and constructing the building.

The "Build" option has traditionally been practiced by large firms that have the resources and experience. No more. Small software firms are getting into the act, urged on by venture capitalists, and facilitated by ethnic ties of the diaspora, as well as cheap communications. We have come across many startups with Indian executives, Israeli executives, and now Chinese executives that build a greenfield site. They draft a friend back home, who leases an office, links the computers, and hires the human resources required. Within a few months, a global technology company is created.

Large firms often hedge on the "Buy-versus-Build" decision and choose both. For example, Cisco, in its India operations, has both captive centers ("Build") and provider relationships with HCL, Infosys, and Wipro ("Buy"). GE, one of the largest foreign players in India, described later in this chapter, has similarly built a portfolio of units inside and outside its corporate boundaries.

Hybrids and partnerships

The pure "Buy" and pure "Build" strategies are both imperfect in some respects, leading many companies to experiment with countless hybrid approaches. Hybrids are used for a number of compelling reasons: in order to reduce the risks for one or both of the partners; when one or both lack resources (such as human resources); in order to accelerate development; or, in order to create a long-term relationship. (We return to these themes in Chapter 6, devoted to legal issues.)

The principal offshore hybrids are presented here.

Joint venture

The partners in a joint venture create a separate, independent legal entity in which both have equity. In the case of offshore IT work, the foreign firm usually finds a local firm as a partner. The local firm is familiar with the local conditions, has experience with local recruitment, and can set up operations relatively quickly. For example, Indian-based Zensar Technologies entered the Chinese market by forming a joint venture with Shenzhen-based Broadengate Systems. A joint venture is a good structure for sharing the financial risks and rewards between the two sides. The disadvantage in this kind of marriage is that each partner comes with different objectives and different organizational cultures.

Alliance

An alliance is also created when partners combine some resources and capabilities, but not in a separate entity like a joint venture. For example, IBM and Indian firm i-flex entered into an alliance, in 2003, in which i-flex develops its banking products on the IBM platform, Websphere. On its part, IBM committed to help market i-flex products. Alliances tend to be fuzzier than joint ventures. Large global corporations tend to have hundreds (sometimes thousands) of alliances, some of which are labeled strategic (because they strive for some kind of competitive advantage) while most are not considered strategic. Companies are reluctant to transfer leading-edge projects to the alliance. In one study, only 13% considered alliances as a significant "technology source."[10] Alliances are less common in offshoring than other hybrids.

Offshore development center (ODC)

This hybrid model is a "pure" offshore era distillation. An offshore development center is a dedicated offshore center established by a provider dedicated to one client, where the client may supply some of the specialized hardware and software. Even though the ODC may be owned by the provider; the workforce, security, and other resources are often segregated just for the one customer. Also, slack resources may be taken up by the client, not by the provider.

Turn-key also known as Build-Operate-Transfer (BOT)

In this hybrid, the client hires a provider to set up the development center and get it running with the intention of taking full ownership after 3 years or more. The client firm is making a strategic choice to reduce its risks by building organizational experience and capabilities first before taking over the offshore center. The client also benefits from faster setup time versus the 12–18 months to build it own "captive" subsidiary. In exchange for the risk reduction provided by the provider, the client is paying a higher price. BOT has been popular in India; for example, Aviva, a UK insurer, wanted to ramp up its IT-enabled services quickly in 2004 and set up three different BOTs with three different Indian providers. Recently, Russian firms have entered this area. For example, Cadence, the large US semiconductor design software firm, contracted with one of the major Russian firms, Mirantis, to establish its Moscow R&D center using BOT.

Staff augmentation

In American business jargon this is a form of contracting for temporary services. The provider supplies labor and manages the standard human resource issues of recruitment, training, and benefits. When offshore companies supply staff augmentation, the workers tend to be foreign staff, often from India, who fly to the client's site using a special work visa, and work at the client site for several months. The client often pays less for this foreign staff than for local professionals. Staff augmentation can be a transitional approach to allow the client to build its experience with the provider, transfer knowledge to the provider's personnel, and generally become more comfortable with offshoring. In practice, some client firms rely on staff augmentation for extended periods of time, often years. This has led to some controversy. Such usage of cheap foreign labor has been the subject of some derision, sometimes called *body shopping*. US-based professor Ron Hira, who has spoken critically of some offshoring practices, calls staff augmentation, tongue-in-cheek, "onsite offshore outsourcing."

In summary, what do most firms prefer: Buy, Build, or Hybrid?[11] We have seen some questionable data on this question. Nevertheless, it is safe to say that the majority of those offshoring are outsourcing, therefore they are using Buy. A useful benchmark is to look at the broader picture of global R&D including manufacturing, pharmaceuticals, and other industries. Companies spend 15–25% of their overall R&D budget on technology outsourcing while only 5–10% on hybrid collaborations (such as joint ventures and alliances), although the value may be higher for hybrids because of their strategic importance.[12]

Multiple providers versus provider partnership

One of the perils of outsourcing is that the client may become too dependent on one major supplier and may be "taken hostage." In offshore outsourcing, in which there are thousands of eager providers to choose from, foreign clients have tried to capitalize on their bargaining power by contracting with more than one provider. This multi-provider approach also coincided with the experimentation stage that many firms have journeyed through (Stage 2 of the Offshore Stage Model depicted earlier in this chapter).

Generally, it is the larger companies which offshore to a cadre of two, or three, or several offshore providers.[13] The managers dealing with these providers become deal-makers, squeezing their suppliers and bargaining for lower bids. More importantly, clients can threaten providers with termination, since the switching costs are low, given that several other providers are already entrenched in the client's business and can quickly take over the slack.

In the early outsourcing era, clients were more likely to view their relationships with suppliers as a "win–lose" relationship. That is, they tried to maximize their benefits from low bids without taking into account the provider perspective. The problem with such an extreme win–lose view stems from the issues discussed earlier in this chapter, namely, that the offshore activities need to go beyond mere cost reduction.

In this respect, both clients and major Indian providers are becoming more astute. The offshore outsourcing relationship has come to be viewed as a partnership. The label "partnership," rather than the more traditional labels of vendor/supplier/provider, has taken on a mantra-like significance in the offshore outsourcing business. Beyond the slogans, partnerships can be melded by creative financial agreements. The first is the increase in creative incentive-based contracts. These are performance-based contracts that give the provider a stake in the client's business outcome (the Balanced Scorecard in Chapter 7 addresses this). The second financial mechanism is an equity holding by one partner in another. For example, Indian-based provider Tata Consultancy Services (TCS) has even invested financially in several of its own clients.

Case study   GE in India

"India's treasure is its intellectual capital,"
                                        Jeff Immelt, Chairman and CEO of GE[14]

Among major US firms, GE has been an offshore trend-setter.[15] While large technology firms, such as IBM and Intel, have been more active in offshoring their software work, GE, a more diversified firm, is an unusual case. This American conglomerate is one of the largest corporations in the world, with sales in 2003 of 133 billion USD and a 2004 market capitalization of 340 billion USD. GE has 31 separate business divisions, in services, manufacturing, and high technology, each of which is required to be one of the biggest in its industry. GE has been particularly successful in its global immersive strategy, in which its international units cross-pollinate each other in supporting cross organizational functions, whether they are in IT, software R&D, back-office functions, manufacturing, or other technology R&D functions, such as materials science.

What really sets GE apart from other multinationals with significant offshore operations is not the specific activities GE performs offshore, but the company's multi-faceted strategic intent. Particularly in India, GE has leveraged not just the country's low labor costs, but also the talents available in its workforce. GE found India's workforce easy to integrate into the global organization – a result of the combination of India's strong education system, British colonial heritage, and widespread fluency in English.

GE was earlier and more aggressive in globalizing its internal operations (including IT) than most large firms. Jack Welch, the revered former CEO, is said to have established the 70:70:70 rule; 70% of all IT work should be outsourced, of which 70% is to be outsourced to preferred providers, and of that, 70% should be at the provider premises to reduce costs. It seems that many in the offshore industry replaced the last clause with "in India".

The 70:70:70 rule became part of GE's performance review and became embedded in GE culture. This goal has been copied, emulated, and pointed to by many managers in the years since then. Even more important, GE's executives and managers, who learned the offshore strategy under the tutelage of Jack Welch, have been hired into influential positions at other important US firms, and have implemented aggressive offshoring strategies at these firms. Kathy Lane, former IT chief for GE's oil and gas division, became CIO at Gillette, a US consumer goods manufacturer. At her first managerial meeting at her new firm, she inquired about the company's offshore strategy. When the response was that there was no offshore strategy, she reportedly used strong language to show her displeasure. Gary Wendt, former head of GE Capital, founded a dot.com era insurance firm and set up all back-office operations in India. While the insurance firm itself collapsed, the back-office operations were efficient and productive enough to stand on their own, and eventually became EXL, a successful IT-enabled services firm.

GE's odyssey is particularly noteworthy in India. As GEers like to tell it, they have been in country for a century. GE entered India in 1902 to build India's first hydroelectric plant. But the period of interest begins much later: in the late 1980s GE began to grow its operations in the country, declaring in 1992 that India was "a priority country." By the mid-1990s GE had several divisions with operations in the country (services, manufacturing, R&D), making everything from medical diagnostic systems to fan belts. Today the majority of GE's businesses have a presence in India, either through a joint venture, a wholly owned subsidiary, a strategic alliance, or a business development and customer support presence.

The John Welch Technology Center was established in Bangalore in 2000 with a facilities investment of 80 million USD including state-of-the-art laboratories and recreational facilities, all in a university-like campus. The center is GE's second largest R&D center and the largest outside the US. Its director reports to the Vice President of

Corporate Research. In 2003, it housed 1800 engineers and scientists, performing tasks for the entire range of GE's businesses. The center generated 95 patents within 3 years of its opening. Originally envisioned to perform R&D only for GE Plastics, the mission was later expanded to many areas, including advanced mechanical engineering, materials, imaging, micro- and nano-structures, chemical engineering and modeling, polymers and synthetic materials, e-engineering, IT, and e-commerce. "This was the natural location for us to go outside the US," said Scott Bayman, President and CEO of GE India. "It is very difficult to get this kind of critical mass in any other country."[16]

IT offshoring evolved in several directions in the Buy-versus-Build continuum. The first was large-scale outsourcing to three of the major Indian providers, TCS, Satyam, and Patni (GE has a 7% stake in Patni). Each of these firms provided GE with dedicated centers labeled Global Development Centers and Global Engineering Centers. These were situated in India's major cities: Mumbai, Bangalore, Hyderabad, Delhi, and Chennai. In each of these centers, the providers maintain separate buildings, often built specifically for GE. In each there are separate security systems and separate networks. In each there are provider employees tasked full-time to GE activities. All of these centers report to GE's global CIO.

GE has experimented with various collaboration strategies over the years. For example, part of the relationship with Satyam was a joint venture, formed in 1998 to provide engineering design, software development, and system maintenance. In 2003, Satyam sold its interest for 4 million USD after GE exercised its option to purchase. In parallel, GE began to build its own greenfield IT centers at three Indian

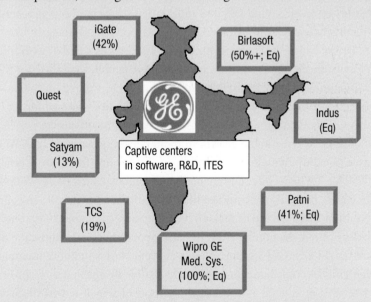

Figure 5.4 GE's centers and partnerships with Indian providers. Percent indicates GE's share of the provider's revenues where known; Eq indicates a GE equity-based partnership.

locations, Bangalore, Delhi, and Hyderabad. The combination of GE-owned centers in addition to multiple suppliers has helped GE's bargaining power. Every 3 years GE renegotiates contracts and squeezes its many providers on costs.

In total, GE-India software activities grew tenfold from 600 software developers in 1995, to 6500 software developers in 2002, and on to 8000 software professionals by 2003. By 2003, half of GE's IT staff were based in India, some of them working in captive GE centers run by the third-party providers. The IT helpdesk with 800 employees was also based in India. GE claims that its offshore IT operations have attained such high levels of quality that some customers are not even doing acceptance testing. Driving all this was cost: GE was relentless in cutting costs in its IT operations, lowering its IT budget to a relatively low 2.5% of revenues in 2003. The reported cost savings in 2002 were 600 million USD. Gary Reiner, GE CIO at the time, was said to be "bullish" on offshore.

One of GE's largest divisions, GE Capital, also began building and growing its ITES in India. From its establishment in 1997, GE Capital International Services (GECIS) has grown to become the largest shared services[17] environment in India, with revenues of 350 million USD. The services provided to other GE divisions and external customers include everything from high-value IT services like ERP and Oracle database consulting, IT helpdesks, data mining and modeling, to consumer and commercial collections, and even insurance underwriting services. By early 2004, these IT-enabled Services centers housed 12,000 workers in eight centers distributed between Gurgaon, Hyderabad, Bangalore, and Jaipur. Its operations were estimated to be saving GE 340 million USD a year. Savings went beyond operational efficiencies and included undisclosed savings from cash flow improvement.

By the end of 2003, an estimated 10–12% of GE's global workforce was in India (if outsourced units' staff are included). GE alone accounted for 8% of all Indian software exports in 2003.[18] By mid-2004, a GE executive, Steve Morrison, GE's director for Global Delivery Centers, recast the now-famous 70:70:70 rule and said that GE should target 80–85% of its IT work to be offshored.[19]

Yet, GE's remarkable growth in India may be slowing.[20] Jack Welch's successor as CEO, Jeff Immelt was focused on China, not India. In a lyrical response to Welch's 70:70:70 rule, GE created an ambitious China policy called the "three 5s" in which, by 2005, the firm would have 5 billion USD in sales in country and source 5 billion USD. On top of that, GE became more concerned with business continuity risks in India (stemming from the new geopolitical risk landscape after September 2001). Finally, GECIS, one of GE's success stories in India, was said to be bloated, with higher costs than more agile independent IT-enabled services firms in India. As a result GE entered into a lengthy period in which it was discussing selling off GECIS as a whole, or in pieces. This culminated in late 2004 with the announcement that GE will sell 60% of GECIS to a US-based consortium.

Case lessons

GE's offshore cost savings have been staggering. But, unlike many other firms that have taken the passage to India, GE leveraged offshoring strategically. GE utilized several of the strategies noted earlier in the chapter. Most importantly, GE's approach was a "global immersive strategy" in which sourcing of IT was but one part of a broad menu of knowledge and service activities. The company seized on the availability of engineers and scientists in India in a wide variety of specific disciplines to build its large ODC as well as its second largest R&D facility. It diversified its technology base globally using its Indian centers. It integrated its Indian centers with those of the rest of the organization. It also fused its many knowledge-based functions successfully: software, R&D, and ITES. Since the 1990s, GE has taken activities that many companies would view as non-core, back-office functions and rolled them into world-class operations in India that set the standards in efficiency and processes for the entire corporation. GE used offshoring as a justification to re-engineer processes far away – to start from scratch with a fresh approach and less political hindrances. Such re-engineering is more difficult to do in the US because costs are much higher. Finally, GE has gained new revenues from offering its services to third parties, from selling various units, and from equity relationships that it had built in India.

Concluding lessons

- Align your offshore strategy with broader corporate business strategies and global sourcing strategies.
- Numeric goals for "how much to offshore" may be useful as incremental goals in achieving a cost-reduction strategy.
- Examine the six unique strategic goals for leveraging offshoring that go beyond mere cost reduction: attaining speed, agility, and flexibility; using talent for innovation; building global networks for knowledge sharing; technological diversification; deeper localization; and new revenue generation.
- Keep in mind the three strategic perils of offshoring: losing core competency; forgetting the broader strategic goals; and losing advantages in proprietary knowledge and code. And don't forget the risks and extra costs discussed in Chapter 2.
- The *Buy-versus-Build-versus-Hybrid* continuum now offers more choices for strategic collaboration. Particularly, choosing the *Build* option, once in the purview of large firms is now accessible to small firms as well.

6 Offshore legal issues

By Rebecca Eisner*

Before you learn to play a new game, you probably ask some questions. What are the rules? What is the objective of the game? How do you win? What are the best strategies? How do you avoid being thrown out of the game?

Business is like a game. Every business person who embarks on a new business deal needs to ask questions to get the information they need to make the right decisions. Some of the questions will be about business and financial issues, and others will be about the rules. You need to understand the rules, and how to play within them. You also need to understand the key risks, and the solutions to mitigate those risks.

In this chapter, we begin with an introduction of seven key legal areas where rules affect your offshore business. Next, the discussion moves on to deal structures, risks and risk mitigation through contract provisions. This chapter focuses on "what" you should be concerned about, not necessarily "how" you can address it. "How" you play the game and play within the rules is beyond our scope.[1]

Key legal considerations in offshoring

Seven of the key offshore legal issues are as follows: intellectual property (IP) protection; labor and employment rights; export control restrictions; privacy and data transfer restrictions; government approval of offshoring; taxes; and currency conversion exposure.

IP protection

The laws protecting IP rights vary from country to country. They are like a patchwork quilt, with holes. This makes it hard to protect IP rights in offshoring arrangements. When software and materials are written in the offshore country, and then sent around the world for use or commercialization, the laws of several countries may apply. Regrettably, the patchwork quilt of international laws regarding IP can leave you cold and exposed.

Software that is written in India is subject to different laws than software that is written in the US. For example, software and business method patents are not recognized in many jurisdictions, such as India, Russia, and China. Rules governing ownership in developed technology and "works made for hire" often differ. In the US, a customer may rely on

* Eisner is an attorney at Mayer, Brown, Rowe & Maw LLP, USA.

the "work made for hire" doctrine to become the exclusive owner of developed technology, but under UK law, no similar right exists. Some countries do not protect trade secrets (e.g. unlike the US, where the Industrial Espionage Act which makes it a criminal offense to steal trade secrets, there is no statutory protection against theft of trade secrets/confidential information in India). Software piracy is rampant in most offshore countries.

Some international treaties and conventions, such as World Trade Organization–Trade-Related aspects of Intellectual Property Rights (WTO–TRIPs) Agreement, and the Paris and Berne Conventions, establish minimum standards for IP protection in their member states. The degree to which governments enforce these standards varies widely.

Most businesses rely on a combination of law and contract to protect their IP. However, contract terms that work well in your country may not be enforced in the same way in another country. There are no guarantees that your IP will be protected, even with great contract provisions. Good contract terms help, but they are not the end of the story.

Consider the following example:

A UK manufacturer of consumer products, called Royal, engages an Indian software company, called Ajit Software Writers, to create custom software for Royal. Ajit has a sales office in London, but no other operations or assets there. Royal provides its software development agreement to Ajit, and Ajit signs. The agreement states that all software written under contract for Royal is to be owned by Royal. Ajit has no license rights to use the software, and the agreement is governed by UK law. As is the typical practice with Ajit, an employee located at Royal's offices in London gathers the high-level requirements for the software, and the software is developed entirely in India by Ajit personnel.

Three years pass and Royal learns that Ajit is licensing the software in India. At this point, Royal's enforcement of its rights in the software becomes quite complicated. Royal may first attempt to enforce its contract by bringing legal action against Ajit in the UK. Royal will want money damages (e.g. lost license fees) and will want Ajit to cease use of the software. If Ajit has sufficient assets in the UK, Royal may get money damages. If Ajit does not have sufficient assets, Royal may be forced to bring actions in India to receive compensation. But money compensation is only part of the story. What Royal really wants is for Ajit to stop providing the software to others. Royal may get the UK court to rule that Ajit must stop licensing the software in India. The UK court's ability to enforce that ruling, however, may be limited, especially if Ajit's business presence in the UK is still limited to a sales office. If Ajit continues to use the software in India, Royal may be forced to go to India to protect its software. Enforcement of Royal's rights may take years.

Enforcement of contract rights in a foreign country can be very difficult. The local courts and enforcement authorities are likely to apply local laws and favor the local company (i.e. Ajit) over the foreign company. Different use rights, registration requirements for protection of IP and laws apply. Royal may spend a lot of time and money attempting to get compensation and to end Ajit's wrongful use of the software. There is no guarantee that Royal will succeed at either objective. In the right cases, international arbitration may be a better dispute resolution process for international disputes.

Labor and employment rights

Labor and employment laws play a huge part in the cost and practicality of offshoring. Most countries have complex laws protecting workers' rights. They can affect not only the customer's employees, but those of the provider as well. You should understand the labor and employment issues in offshoring, especially where existing workers' jobs may be eliminated or replaced. For example, in the US, the US WARN Act requires employers with 100 or more employees to provide a 60-day notice to displaced workers in certain circumstances.

In the European Union (EU), countries have enacted regulations that implement the Acquired Rights Directive to protect workers' rights. The purpose of the Acquired Rights Directive is to safeguard the rights of employees on the transfer of a business or a portion of a business to a new employer. Under certain circumstances, the Directive and country regulations may apply to offshoring and/or outsourcing.

If the Directive applies, then the transfer or elimination of employees must comply with various principles and regulations. These principles and regulations may require that employees transfer with the entity, and that they get the same terms and conditions that they enjoyed immediately prior to the transfer. Employees who are dismissed before or after the transfer may have claims for unfair dismissal. Finally, representatives of the employees affected by the transfer are entitled to be informed about the transfer and to be consulted on measures which are proposed as a result of the transfer.

In offshoring, employees often do not transfer to the offshore provider. However, the Directive may still apply to the affected employees. In such case, the employees may be entitled to notice and counseling, if they are affected by job replacement, changes, or release due to the offshoring. Severance payments and other rights may also be triggered.

Other countries impose stringent hiring and firing requirements on employers, requiring government approval for layoffs and closures. For example, in India, if a company employs over 100 employees, a government approval may be required to fire employees. If such approval is denied, the company may be forced to implement voluntary retirement schemes and pay the employees to resign. In Canada, some Eastern European and some Latin American countries, workers who are transitioned to an outsourcing provider may be entitled to severance payments even though they obtain a new position with the outsourcing provider.

Immigration laws also affect offshoring, particularly where foreign workers need to be close to the customer. Under US immigration law, in September of 2003, the number of H1-B work permits available was reduced from 195,000 to 65,000 annually. This reduction of H1-B work permits may limit an offshore provider's ability to place staff onsite in the US. Outsourcing agreements should allocate responsibility to the provider for managing such visa, immigration and qualification to work issues. It is important for the customer to know that provider personnel hold all appropriate work permits and authorizations required under applicable laws.

Import and export issues

Various countries have export restrictions that affect the types of products and services that may be sent across borders. For example, certain types of software may not be sent to or from various countries. In the US, certain software products with strong encryption capabilities are regulated as weapons by the Department of Defense. It is illegal to transport this software outside of the US without first obtaining the appropriate approvals. Thus, a company in the US that desires to send regulated software offshore for maintenance and development may not be able to do so without the appropriate approvals and licenses, or may not be able to do so at all. Similarly, China regulates the import of encryption software.

Countries often have laws that restrict access to certain types of sensitive business and government data. In the US, a company that sells products and services to the federal government may have sensitive data that is protected by regulations and agreements with the federal government. In those cases, the sensitive data which may be part of a larger database, may only be accessed and used by approved individuals in the company. Access to such data, by foreign citizens inside or out of the company may be a violation of the regulations and agreements.

Privacy and data transfer

Recent developments in privacy laws worldwide have created some complications for offshoring transactions. Privacy and its close cousin, data security, are emerging as key new topics that present both legal and business risks. Failure to consider and plan for privacy issues can bring unwanted consequences, such as bad publicity, official enforcement actions, fines and penalties, and private lawsuits. Even more damaging is the loss of public trust that can result from privacy problems.

US privacy landscape

The US historically has favored self-regulation for privacy protections. This meant that, until recently, there was little US privacy law to consider in offshoring. Technology has brought big changes – robust databases, data mining, CRM tools, cookies, cross-matching of data, Internet use, data sharing, offshoring and outsourcing. These changes are seen as a threat to privacy rights. Consumer protection groups and governments are expressing privacy concerns.

The US Congress enacted the following privacy legislation in the areas of personal financial, health and medical information:

- The *Gramm-Leach-Bliley Act* governs personal financial information.
- *Health Insurance Portability and Accountability Act (HIPAA)* covers health and medical information.

- *Children's Online Privacy Protection Act (COPPA)* governs information collected online from children under the age of 13.

Apart from the recent laws and a few prior existing ones, many US businesses rely on self-regulation, including voluntary industry guidelines, membership in privacy certification programs, such as TRUSTe, or compliance with a self-established privacy statement and program. The Federal Trade Commission has taken an increasingly active role in privacy matters.

States, too, are beginning to add to the growing body of privacy law and regulation. The State of California has a tough data security law (effective July of 2003) that requires notification to individuals of possible security breaches involving the compromise of their personal data. Other states may follow California's lead.

European Union Data Privacy Directive

The EU has been a leader in enacting and enforcing privacy regulation. Companies that collect or process data in the EU, or that receive data from the EU, are most likely subject to the EU privacy regulations.

The European Union Data Privacy Directive (95/46/EU) was adopted by the European Commission in 1995. It required the EU member states to enact legislation in accordance with the Directive by late 1998. The Directive has been implemented in EU countries through this national legislation. For our purposes, references to EU privacy law means, both the Directive and the national legislation.

The EU privacy law applies to any business that collects and processes personal data on EU residents. You do not have to be located in an EU country to be subject to the EU privacy laws. The EU privacy laws regulate the collection and processing of employee data, customer data, patient data, and other personal information. This affects many areas of outsourcing, such as the outsourcing of human resource functions, financial functions, and IT functions where personal data is involved.

A particularly critical area covered by EU privacy laws is the transfer of data to countries outside the EU, even if you (or your offshore provider) are just transferring your own internal data from your EU operation to another operation. The EU privacy laws limit the export of regulated data to countries that do not offer "adequate protection". Only a few countries that are not members of the EU have been approved by the EU to receive this regulated data. They include Switzerland, and Canada. India, China, Malaysia, the Philippines, Russia, and many other offshoring destinations are not yet deemed to have adequate protection. You cannot send EU personal data to these countries unless you use one of the approved methods of transfer.

There are several ways to accomplish these transfers legally, but none of them are easy. For example, if you want to transfer data on your customers in France to India, you could ask the French data privacy authorities for approval of the transfer. This could be time consuming, and it may require you to keep going back to those authorities for approval as facts or circumstances change. Alternatively, you could get each customer's consent, which could be an onerous task. In an offshoring context, the most efficient way to handle data transfers is likely through use of EU approved data transfer contract clauses. These clauses require the data transferor and transferee to agree to a set of contract provisions that are consistent with the data privacy laws, and that allow enforcement by the EU authorities and the people that the data describe, referred to in the EU privacy law as data subjects. Another alternative available for transfers of data to the US, may be the US "Safe Harbor" arrangements. Under the "Safe Harbor" scheme, a US company may self-certify to the US Department of Justice that the company is in compliance with the Safe Harbor data protection principles. The EU has agreed to permit transfers of EU data to companies that have self-certified as to their Safe Harbor status.

Failure to comply may result in enforcement actions by the EU authorities in various EU countries. Each country has the ability to enact and enforce its own sanctions. In the UK, individuals and/or corporate bodies may be prosecuted and fined for violations. Spain's laws carry high fines, up to 600,000 USD per violation. Spain has already pursued two well-known companies (Microsoft and Telefonica) for violation of its data laws. In addition to fines, enforcement actions can include interruption or shutdown of your data collection, data processing and data transfers. Failure to comply can also bring on private lawsuits from data subjects. All of these things can seriously damage the reputation of your company or business.

Other international developments

Many other countries are following the EU's lead in regulating data privacy. Some countries that would like to gain admission to the EU are considering laws similar to those of the EU. Other countries such as Canada, Australia, Argentina, and Japan have enacted or are considering their own new data privacy laws. India has not yet passed data protection and privacy measures similar to those in the EU, although India is considering entering into an arrangement with the EU that would provide a means for data to be transferred from the EU to India.

Offshoring and data privacy compliance

Privacy laws generally put the burden of compliance on the client, not the provider. If you are the client, consider the following suggestions for managing privacy issues with your provider:

- *Know the Privacy Laws, but make sure your provider knows them too.* Usually the customer will shoulder most of the direct obligations under the privacy laws. Outsourcing providers will seek to shift responsibility and cost for tracking new developments in the law to the client. This shift may not be appropriate in all cases, especially when the provider has multiple clients who are subject to privacy laws. You will need to negotiate the proper allocation of responsibility for staying up to date on the changing privacy laws.

- *Don't pay the provider's whole tab for compliance*. Compliance with privacy laws costs money. You and the provider may have to consider changes in technology infrastructure, data handling procedures, security measures, data storage, locations of data centers, information sharing policies, and many others. A good offshore provider will already be familiar with the laws applicable to it and its clients and will have taken action to comply. Resist provider attempts to present you with the whole bill for compliance.
- *Get strong contractual assurances*. Your outsourcing provider should agree to a variety of provisions aimed at helping you to comply with privacy laws. These include your control over and access to the data; the use of appropriate data security measures; restrictions on data use, transfer, processing, and sharing; an agreement to make changes as required by changes in privacy laws; facility audit rights; and many other similar topics. In some cases, the privacy laws may require use of specific contractual provisions, as is the case with the EU privacy laws.

Government approval of outsourcing

Offshoring is a hot political topic in Europe and the US, as discussed in Chapter 12, Offshore Politics. By 2004, in the US, the federal government and more than 30 states had considered legislation to limit offshoring. These proposed laws reflect a growing trend toward regulation of offshoring. Some of the proposed measures require that prior notice be given to affected employees. Others seek to prohibit offshore outsourcing altogether. Similar developments have been taking place in Europe. Rapid developments in this area mean that organizations considering an offshore outsourcing arrangement have no choice but to monitor these developments.

Aside from laws seeking to regulate offshoring, political relations with the target country can be a factor as well. Political discord can lead to regulations that impact the offshore services, such as the imposition of quotas, taxes and tariffs, restrictions on foreign ownership or control, embargoes and other similar measures. In 1998, the US imposed sanctions against both India and Pakistan for nuclear testing. While these sanctions did not directly affect offshore outsourcing arrangements, sudden actions like an embargo may disrupt offshoring.

Taxes

Offshoring services to a third party can have a tax impact. Applicable service taxes must be considered to have a complete picture of the cost and potential savings of offshoring. In addition, the contract and deal structure should provide for the minimization or recovery of such taxes to the extent legally possible.

Some taxing authorities, including some states in the US, impose taxes on the provision of services. For example, when a company provides services for itself at a location in the State of Texas, US, the company does not incur any service taxes in connection with those services. If the company outsources those services (whether domestically or offshore), the services that are provided to the Texas location may be subject to a services tax. Withholding taxes may apply in international transactions, and many EU and other countries have VAT taxes which apply to goods and services provided or sold within those countries.

Currency

If payment will be made in the currency of one country and converted into the currency of another country, there is the issue of fluctuating currency conversion rates and the risk that one or both parties take regarding the relative strength or weakness of their currencies. Also, there is the risk that it may be costly or impossible to convert currency at all. Some governments recognize particular currencies for conversion and reporting purposes. For example, the Chinese government regulates the flow of foreign currencies in and out of China, and requires certain documentation evidencing the underlying transaction. China also dictates the exchange rate and restricts use of Chinese currency to pay obligations to foreign entities.

Sometimes it is advantageous to fix the particular currency of payment and the conversion rate so that both parties understand the nature of the currency risk going into the deal. This gives either party the opportunity to hedge or correct for that market risk. Alternatively, it may be advantageous to let the currency conversion rates float with the market. It may also be advantageous to allow a party to dictate payment in a convertible currency if regulations reduce convertability. In any event, it is important for both customers and providers to understand the risk associated with payment in foreign currencies, and in particular, whether such risks create any additional costs.

Principal deal structures

Four basic deal structures are used when offshoring. Each of these four principal structures comes with various benefits and burdens. In some cases they involve balancing the risk mitigation strategy against the anticipated benefits of the offshoring. In Chapter 5, Offshore Strategy, these deal structures were discussed from a collaborative strategy perspective, but the risks and benefits are worth re-visiting from a legal perspective.

Captive center/subsidiary

Some companies create their own offshore service and development centers. Companies that use this option must comply with local laws. For example, establishing a new business location generally requires registration with various authorities (federal, provincial, state, local, and often municipal). There may also be local corporation laws that dictate who may own the company and according to what ownership structure and interests, who may control the board or management committee of the company, and a host of other corporate governance considerations.

Special permits and operating licenses may be needed. Local employment laws may impose wage and benefit requirements, collective bargaining agreements, and other similar requirements. There can also be significant tax implications to establishing a captive offshore facility. This is because returns from the captive entity may be subject to tax in the offshore location as well as in the parent company's location.

Joint venture

In this alternative, the customer and the provider form and own a joint venture in the offshore country. The joint venture then services the needs of the customer. The joint venture may also sell services to third parties.

A joint venture can align incentives and goals of the customer and the provider, in part through sharing of profits and losses. However, joint ventures can be complex to establish and govern. They require initial investment, and they may be expensive to exit. In addition, the parties establishing the joint venture still need to comply with local laws for establishing and running the business, and they deal with the same issues that companies have when they establish their own captive presence in the offshore location. As with captive organizations, the tax implications of a joint venture must be considered in looking at the total cost of the option.

Build Operate Transfer

In this model, the customer hires a provider to build and operate a service organization, with an option or the obligation to purchase the established entity after a certain period of time. A Build Operate Transfer (BOT) model may require a low to moderate initial investment. Issues of legal compliance with local laws are generally left to the service provider until the transfer occurs. The BOT model allows the customer to become familiar with the legal requirements over time and well before the customer takes control of the BOT service organization. In addition, since BOT models are usually viewed as a service arrangement, they are unlikely to have some of the same tax disadvantages that may exist with captive or joint venture models, at least prior to ownership of the BOT organization.

Outsourcing

A great number of offshoring arrangements are completed through the traditional customer and provider services agreement. In a traditional services arrangement, the customer transfers responsibility for certain services to the provider. Often this process is started with a request for proposal (RFP) to one or more providers. The providers respond describing their service delivery solutions, their capabilities and their pricing, and the customer selects one or more providers with whom to negotiate offshore agreements.

Traditional offshore outsourcing arrangements have various advantages and disadvantages over captive arrangements and joint ventures, and in some cases, BOT models. The opportunity to lower costs due to a competitive bidding structure is a major advantage. Providers have a greater ability to maximize efficiencies and lower costs because they are typically servicing many customers. Providers can usually provide higher service levels due to their specialization and efficiencies. The disadvantages may include:

- loss of control,
- loss of flexibility,
- the possibility of misaligned incentives between the customer and the provider,
- cost overruns.

Solid contractual provisions can secure some of the advantages and mitigate some of the key disadvantages in offshoring to a provider. There are also different contract structures that can minimize the risks as well. These structures and contractual terms are discussed below in the next section.

Agreement structures

Offshore outsourcing agreements take many forms. Below are some common approaches used to offshore software development activity.

Common approaches

Pilots. Pilots are a means of testing offshore outsourcing. For example, a company may decide to offshore maintenance for a limited set of non-core applications, and gradually increase the scope to more critical applications, if the offshore arrangement proves successful. A major American technology company used multiple pilot projects to begin work with providers in India, and China. Through trial and error the company learned that certain providers are better at providing resources for small jobs, while others are more suited to longer and more complex ones. The company then expanded its outsourcing relationships with those companies that had performed well in the pilot phases.

Short terms. Other companies commit the particular scope of the function or service up front, but in an agreement with a short term, subject to options to extend. These agreements contain more detail than a pilot program agreement based on the assumption that the term will be extended.

Full-scale outsourcing. A third model is a more robust and defined outsourcing arrangement, with a large defined list of services in scope, a detailed plan for transition of the work to the service provider, detailed service level agreements (SLA), governance and relationship management provisions, policies and procedures manuals, and many other typical outsourcing agreement terms. A global semiconductor company used a pilot arrangement to assess an Indian provider's capabilities and then, when the pilot succeeded, moved to full-scale outsourcing.

Multiple suppliers. Some companies prefer to have multiple offshore providers to reduce reliance on any one provider, to maintain competitive pressure and to create flexibility. This approach works particularly well for project work that can easily be reallocated. It can reduce country risk if the providers are in different locations. It can also provide a disaster-recovery alternative.

Using multiple providers also has its disadvantages. Multiple providers means double or triple the governance, management and coordination. Use of multiple providers also complicates determining who is responsible when defects are detected. It may also result in less favorable contract terms and higher prices than a company might receive if it is willing to pool all of its business with one provider.

Choosing the right partner: financial stability and location

The financial ability of an offshore provider to perform its obligations is a critical consideration, particularly if you will depend on a single provider for a critical service. You must evaluate the risks associated with a provider and consider a few questions: Who will be the contracting party? Where are they based? Are they thinly capitalized? Do they have substantial assets and revenues? How much risk is your company willing to take in contracting with a provider who might not be well capitalized, or who may not have the performance record of some of the other providers?

Where a provider is thinly capitalized, a customer may require a guarantee, letter of credit or other type of security for performance and payments. Alternatively, the contract should be structured to minimize payment and performance risks, and the pricing should reflect the degree of risk that the customer is taking.

In addition to the financial capability of the provider, the location of the provider's assets and business may determine where and how you ultimately enforce your agreement. The ability to enforce contract terms varies from country to country. You should consider which party is the best partner for your company based in part on the degree of risk that you and your company are willing to accept for the offshore project.

Consider the following scenarios:

- *Onshore provider with offshore capabilities.* FirstCo is based in the UK, and desires to offshore certain software development services. FirstCo has operations in Europe, but does not have a significant presence in India or Asia. FirstCo would like to choose a provider that has substantial operations in India. FirstCo is concerned about the risks of contracting directly with an Indian-based provider. FirstCo would rather contract with an established UK service provider, SoftCo, who has offshore capabilities. SoftCo has a well-established presence in the UK, and has sufficient assets against which SoftCo may rely should things under the contract go wrong. FirstCo will require in its contract that SoftCo will do everything necessary to ensure that its offshore affiliates and subcontractors will comply with the contract. FirstCo will require that all contract disputes be handled in the UK under UK law.

- *Offshore provider with onshore capabilities.* SecondCo is a multinational company with operations and locations around the world, and headquarters in the US. SecondCo has manufacturing facilities in India and China, but no established operations that would support software development services. SecondCo currently completes software development through two major centers around the world: one in the US, and other in Europe. SecondCo wants to relocate some of its software development into two centers, one based in China and the other in India. SecondCo is comfortable with contracting with providers in India and China directly because SecondCo already operates there. SecondCo signs an agreement with an Indian-based provider of services who has substantial operations in the US, India, and China. The agreement is governed by US law, and most major issues will be resolved in accordance with US law and in US courts. SecondCo knows that if legal issues arise, it will enforce its agreement first in the US, but it may need to resort to enforcement in India and possibly in China with the provider. SecondCo has the resources and the contacts to take those actions, if necessary. SecondCo is comfortable taking this increased risk. As with FirstCo, SecondCo has performed due diligence on the provider to know that the provider has sufficient assets and operations in the US, India, and China, such that SecondCo may enforce its contract rights first in the US, but also in India or China, if necessary. SecondCo may also consider international arbitration to resolve disputes.
- *Offshore provider, a new entrant.* ThirdCo is a US company that has some international presence, but not as extensive as SecondCo. ThirdCo wants to offshore software development services to India. ThirdCo has identified an Indian service provider, CodeWell. CodeWell has performed well in a pilot program. CodeWell is a subsidiary of a large manufacturing and engineering company based in India whose primary business is construction. CodeWell is a relatively new entrant to the IT services market, and CodeWell only accounts for about 3% of the parent company's combined revenues. ThirdCo is nervous about contracting with CodeWell because CodeWell has relatively few assets. CodeWell's parent company is a much larger company with substantial assets. ThirdCo will require that CodeWell's parent give a guarantee of the financial and performance obligations of CodeWell under the agreement. In this way, ThirdCo is assured that a larger and more substantial company is standing behind the new entrant's obligations. ThirdCo will use international arbitration to enforce the guarantee. The arbitration will occur in a neutral country, which is fair to both parties.

In addition to choosing the right party, there are contract provisions that may be useful in protecting against a financially challenged provider. Make sure your contract gives you ownership in important assets: that is, in the case of software, including source code. Where you will not own the source code, and are using software developed under license from the provider, you should require that the provider's source code be placed in escrow, and that you have a current license grant permitting you to access and use that code. Consider negotiating a right to terminate after a drop in the provider's credit rating, or after the provider suffers a material adverse event. Require the provider to deliver regular financial statements, certified by its chief financial officer.

You should structure payment terms, such that services are received, or milestones are met before you are required to pay. You should not allow the provider to assign your agreement to a different party without your approval. Require the provider to carry insurance and to name you as a loss payee or additional insured. Finally, include clear clauses that require the provider to return your confidential information, owned materials, and other proprietary items at any time upon your request.

All of the clauses above together will not prevent you from suffering business disruption and loss, if your provider suffers financial trouble. But they can lessen the blow.

Key service agreement terms

Some contract terms common to outsourcing deals take on unique importance in offshoring. The contract terms in this section address the key risks and key disadvantages discussed in earlier sections of this chapter regarding, contracting for offshore services. These contract terms apply to more than a traditional offshore outsourcing agreement. They would also apply to the end user of services in a joint venture or BOT arrangement. In a joint venture or BOT situation, there is usually a service agreement that covers the details of the service that the end user requires. The service agreement is separate from the corporate "deal" documents that establish and govern the joint venture or BOT organization. In fact, it is important for the customer to document the services to be provided by the joint venture entity or the provider in a BOT model in the same manner that the customer would if the customer simply hired a provider of services without the joint venture or BOT part of the deal.

Rights to approve personnel and subcontractors

The customer should have a contractual right to approve of all key provider personnel and contractors, and to remove them for non-performance. These rights to approve are important for several reasons:
- The customer retains a level of control over who performs the services.
- The customer may need to approve of subcontractors for legal reasons, where certain functions cannot be further subcontracted or moved to different countries (e.g. an offshore provider could not further subcontract services to a country that is subject to current embargo restriction).
- The fact that much of the work is being performed by provider personnel in distant locations presents a unique contract and quality management challenge. The

remoteness of offshore services adds to the importance of the project management team and onsite staff, all of which need to remain subject to the customer's reasonable approval.

In contrast to domestic outsourcing, in offshoring very few customer employees transfer to the provider. This creates a risk of discontinuity and a potential loss of institutional knowledge. To guard against knowledge loss, the contract should address the mix of onsite, onshore, and offshore staff, including: defining the required qualifications and experience of support personnel; defining the required staffing levels; describing the provider's retention strategies for any of the customer's existing staff who will be retained through a transition period; and defining the processes for ensuring effective knowledge transfer.

Rights to dispute charges

An outsourcing customer should negotiate for the right to dispute charges in good faith, and to withhold payment of those charges until the issues are resolved. In light of the enforcement issues that may exist in offshore outsourcing arrangements discussed earlier in this chapter, this right is terribly important to protect the customer. While this remedy is not a favorite with providers, there are a few measures that providers may request in the contract to reduce the chance that a customer will unfairly withhold payment. A provider may request that payment disputes go through an expedited dispute resolution process, so that the provider is not waiting endlessly for payment. The provider may also require that disputed payments be held in escrow by a third party until the dispute is resolved. Holding money in escrow deprives the customer of use of the funds while the dispute is pending. An escrow protects the provider and encourages the customer to dispute charges only when the customer has a good faith dispute.

Termination rights and unwinding

Unlike other commercial contracts, outsourcing arrangements typically permit the customer to terminate for various reasons, but do not permit the same termination rights to the provider. These customer rights might include: termination for provider non-performance; termination for provider change of control; termination for certain service level failures; and other similar rights. The customer may also be able to terminate without cause, usually with some prior notice and in many cases with payment of a termination charge. The provider may only have the right to terminate for repeated non-payment by the customer.

Outsourcing agreements can protect customers, using extensive provisions regarding what happens upon termination. The unwinding provisions take on even greater importance given the complexity of distance and the difficulty of knowledge transfer from provider personnel back to the customer or the customer's new provider. Usually the customer is entitled to termination assistance or unwinding services from the provider, in addition to other termination rights.

The customer may seek rights to hire dedicated onsite employees and contractors of the provider, and potentially to hire some offshore employees as well, to preserve knowledge. If hiring employees is not possible, then it is important for the customer to have a long termination assistance period to enable knowledge transfer. In addition, customers may have the right to purchase dedicated equipment from the provider, although the customer needs to consider whether decommissioning and transport costs make this right less attractive when the equipment is in an offshore location.

Another important right, upon termination, is to require the provider to assign to the customer any dedicated third-party agreements and licenses. The customer may want to specify certain terms of the dedicated agreements that the provider enters because the customer may want to take assignment of these agreements after termination.

The agreement should also specify that the provider should deliver all work in progress, and return the customer's confidential information.

Finally, there may be software tools and other proprietary materials owned by the provider, and used by the provider to provide the services. The agreement should specify that the provider will license those proprietary items to the customer upon request. This is especially important in offshore deals where many providers have special code creation tools, compilers and other items that they use to provide the services, but they do not commercially license to other parties.

Right to use third parties or in-source

One of the more important ways for a customer to retain flexibility and control is the right of the customer to re-bid the services to a third party or to perform the services internally. This can be the best method for ensuring that the customer is obtaining a competitive price for the service. The mere prospect that this right could be exercised can be an important reminder to the provider of the need to remain competitive in the cost and quality of its services.

Price protections

Pricing for offshore deals varies from fixed price engagements to time and materials work priced against a negotiated rate card. Fixed price engagements tend to be used when the project requirements are well defined, and most of the variables are known. For multi-year offshore agreements, where work is done over time (such as in a large application maintenance and development deal), the prices for the services are often determined according to a rate card. When the customer's costs to receive services are not fixed, it is important to include price protections in the agreement. There is no perfect price protection mechanism. Outsourcing customers should look for a variety of provisions that together work to deliver a market-priced deal.

First, customers should look for the agreement to provide "all inclusive" pricing. Some offshore providers charge additional money for use of certain tools and software. Aside from the base charges for services, it is also important to agree on who will pay the incidentals such as travel, lodging and communications. Many customers accustomed to onshore services forget the additional expense created when offshore professionals travel to the onshore site. Similarly, the parties must decide who will pay for the network security and data communications that will be required for the offshore site to communicate with the onshore facilities.

Second, customers should expect to pay for rate increases in the labor portion of the charge for services. The rates are often subject to increases according to some economic adjustment factor, such as a cost of labor adjustment (COLA), or a consumer price index (CPI). The agreement should reference the right index for these escalations. For example, for a rate card that covers providing software development resources in both Hungary and India, the rate card's escalations factors should be defined by the best index available for each country. Customers may want to cap the maximum amount of these increases, and equally split any increases that exceed the cap.

Finally, we come to the special case where the volume of business is large, typically in a multi-year agreement. Here, the customer has committed to providing large volumes of business to the provider, or the volume of business makes the customer heavily dependent upon the provider's services. In such a case the customer should consider including *benchmarking* rights. In an agreement that includes benchmarking, a qualified third party analyzes the provider's price, service levels, service quality and other factors as compared to similar providers' prices and services. If the provider's prices are determined to be high, the customer may have the right to reduce the services, require a price reduction for future services, or terminate the agreement. Benchmarking for offshore services should be done according to geography, service type, and skill set. For example, a benchmarking that examines the rates for SAP-qualified developers in India from top tier providers will be more accurate than a benchmarking that compares the rates for all software developers in the US, Europe, and India.

Service Level Agreement (SLA)

There are a number of direct and indirect measures of service quality that may be appropriate to include in the contract. They include financial levels and performance milestones for provider compliance. They also include statements regarding compliance with recognized industry standards, such as Capability Maturity Model (CMM) and International Standards Organization (ISO) certifications.

Service levels are perhaps the most important and widely recognized measure of service quality. Defining service levels is critical because service levels represent the objective standard used to measure the provider's performance. The SLA assigns to the customer rights and remedies if the provider fails to achieve specified performance levels. Some critical SLA terms and principals are described below. Another detailed discussion of this important element of offshore outsourcing is found in Chapter 7, in the Governance section.

What is measured?

Service levels are defined ways of measuring a provider's performance. A service level is a measure of the quality, speed, availability, capacity, reliability, user-friendliness, time-liness, conformity, efficiency, or effectiveness of services. For example, an availability service level for a computer system might be the percentage of the relevant time when the computer system is capable of performing a specific task. Service levels for a soft-ware development project might include measuring whether the project is done on time, within budget and within a tolerance for defects and errors.

A good service level is designed to align the incentives of the provider and the cus-tomer. For example, a fixed price contract may incent a provider to cut costs (and qual-ity) in order to increase profits. The SLA for a fixed-price contract should focus on quality and timeliness.

How is it measured?

The parties must define the service level with precision. For example, is a computer system "available" if its central processing unit (CPU) is working? Or do the databases and telecommunications systems also need to be working? Does it need to be "available" to the end user, who may not be able to access that computer system because of a local area network failure? Is it "available" when the operating system is working, even if the application program has failed?

For each service level, you need a process for measuring provider's performance. For example, you could measure a computer system's availability in several different ways: by installing a resident monitoring program within that computer system, through periodic polling by another computer system, by user complaints about downtime or use of a monthly user satisfaction survey asking about perceptions of downtime. The measurement process will affect the results.

You also need to consider the measurement period over which you will measure the service level. Typically, the measurement period will be a month or quarter. Longer meas-urement periods give the provider more opportunity to make up for bad performance. Shorter measurement periods give the provider a "fresh start" more often. Longer meas-urement periods mean that more is at stake during any one measurement period. The measurement period may exclude excused "downtime" due to scheduled maintenance, acts of war or terror, or other events beyond the provider's reasonable control.

SLA reporting

The SLA should require the provider to make performance reports on a timely basis for each measurement period. The SLA should define precisely what information will appear on the reports: such as exception reports for missed service levels and trend reports for key service levels. The SLA might also require the provider to conduct a root-cause analysis of service level failures and report the results to the customer.

SLA credits and performance incentives

A "service level credit" is a credit that the provider grants to the customer after a service level failure. The provider may be required to write a check to the customer or the customer may simply have the right to apply the credit to future service. Either way, it reduces the price of the services and the provider's profit margin.

As an example, an SLA might call for service level credits for any of 10 service levels. For each of those 10 service levels, the SLA might indicate a number of "credits" to be granted upon a failure, with each "credit" being a small percentage of the customer's total bill for the measurement period. The total service level credits for a measurement period might be capped at, say, 10% or 15% of the total monthly bill.

Service level credits are an incentive system. Customers often retain the right to revise the service level credit structure so that they can re-align the incentives as their priorities change. One important question is whether the service level credits are the customer's sole remedy for a breach or merely one of the customer's remedies. The contract should clearly state whether the credits are the sole remedy or are in addition to other contract remedies.

Dispute resolution

Disputes are inevitable. In outsourcing arrangements, particularly offshore outsourcing arrangements, it is critical that the parties have a process in the agreement for quickly handling the many disputes that could arise. This process may include informal discussions between the primary contacts at the customer and the provider, followed by escalations to more senior members of each organization after certain periods of time. If such escalations do not resolve the issue, other more formal actions may be taken, such as binding or non-binding mediation, arbitration, or bringing a claim in the applicable court or tribunal. These formal actions should only be used after informal processes have failed. The parties should not dash off to court every time they have a disagreement. Unlike games, where there is often a winner and a loser, successful outsourcing relationships acknowledge the need for mutual benefit of the provider and customer. This mutual benefit requires that the parties cooperate, and when appropriate, compromise. Formal dispute proceedings, while useful and necessary at times, do not necessarily facilitate cooperation and compromise. The parties should recognize that use of formal dispute resolution may signal the end of the relationship.

THIS CHAPTER IS FOR REFERENCE PURPOSES ONLY. IT IS NOT A COMPREHENSIVE TREATMENT OF ALL LAWS AND REGULATIONS PERTAINING TO OFFSHORE TRANSACTIONS, AND SHOULD NOT REPLACE CONSULTATION WITH LEGAL COUNSEL. THE READER SHOULD SEEK SPECIFIC LEGAL ADVICE BEFORE TAKING ANY ACTION DISCUSSED IN THIS CHAPTER.

7 Managing the offshore transition

Erran Carmel and Erik Beulen*

This chapter covers the management processes and structures that lead to successful transitioning of IT work offshore. These are lengthy, subtle, and difficult processes that require close managerial attention. We cover three key transition topics in this chapter, and each one of these topics can be read by itself and does not require linear-style reading. The topics are augmented with case studies and examples. The three topics are:

- *Knowledge transfer*. The process of successfully transferring specific types of knowledge and experience into the *minds* of all those collaborating on the work across many kilometers.
- *Change management*. Overcoming the inertia and resistance within the organization to a difficult change.
- *Governance*. Establishing structures, roles, responsibilities, and written agreements to ensure that control, coordination, and relationships are all functioning smoothly between the client and the provider in offshore *out*sourcing.

Knowledge transfer

> *How would you transfer the knowledge of eating at a restaurant to someone who never has? You don't really know about it unless you go there yourself. You can have someone to tell you about it. You can order takeout from a restaurant. You can buy a cookbook from a restaurant. But, to really understand how it works and feels and tastes, you have to go to one.[1]*

Knowledge transfer deals with moving specific kinds of knowledge and experience into the minds of the people collaborating on the software work.[2] Some of the knowledge transfer is from home location to offshore location; some of it goes the other way.

Knowledge transfer is one of the principal reasons for failures in the first few years of offshoring, regardless of whether one is dealing with offshoring by outsourcing or offshoring inside the firm. Executives, intoxicated by offshore euphoria, demand a quick payback, leading their subordinates and their providers to take short-cuts. These short-cuts often prove to be expensive.[3]

* Beulen is at Atos Origin and affiliated with Tilburg University, The Netherlands.

The root of this failure is that companies do not manage the patient process of knowledge transfer. And, perhaps the offshore outsourcing provider is pressured to deliver quick returns and reluctantly agrees to overly ambitious transition plans that hinge on rapid knowledge transfer. When complex applications are produced offshore without sufficient attention to knowledge transfer, the problems will be discovered during the costly integration and implementation phases. In such a case, the offshore development costs are, indeed, cheaper, but they are washed away in the final stages of the project life cycle when extra resources are required to correct mistakes. Similar dynamics occur when infrastructure management activities are offshored without patient knowledge transfer. Many issues will revert back to knowledgeable experts at the home location, thus negating the labor cost savings.

Some of the knowledge that needs to be transferred offshore can be codified and written down. This is often called *explicit knowledge*. These are the facts, principles, and specifications. In general, explicit knowledge tends to be the easiest one to transfer. Such knowledge may be captured in Knowledge Management Systems that many organizations have today. But in many organizations much of the knowledge that can be codified is not documented. So, time and effort must be invested to document this knowledge when the offshore engagement begins.

The more difficult knowledge transfer is for *tacit* knowledge.[4] Tacit knowledge is that which is difficult to write, document, or codify. It is fuzzy knowledge learned from practice, exposure, and experience: in other words, the "know-how." It is also the "know-who" of social relations. Much of the tacit knowledge can only be transferred through learning by doing, through "show-how," when one person learns on-the-job through mentoring and coaching.

To use an analogy in the game of chess, the game rules can be documented in just a few pages. These rules represent the explicit knowledge. But this is not sufficient to become a strong chess player. The knowledge to play chess well is the tacit knowledge learned from experience, from trail and error, from coaching, and from very specialized books on chess strategy.

The four types of knowledge that need to be transferred offshore are (see Figure 7.1):

- *Skills*, such as new programming language.
- *Process*, such as harmonizing methodologies between onshore and offshore sites.
- *Domain*, such as business, scientific, algorithmic, and artistic.
- *Work and cultural norms*, such as organizational and national culture.

The first type of knowledge transfer is the least problematic: transferring specific skills, such as new tools, special programming languages, or specialized packages. Much of this knowledge can be transferred in writing. Additionally, classes are offered in most countries, so travel is not required for transfer.

Transferring process knowledge is somewhat more difficult. In essence it is about harmonizing methodologies between the distant units so that they can collaborate effectively. This issue has appeared most prominently in the mismatch between the large Indian

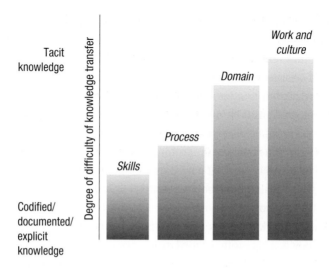

Figure 7.1 Knowledge transfer types.

providers practicing advanced process methodologies (such as the Capability Maturity Model (CMM)) and their clients in which process maturity is low or lacking. In these cases it is the client staff, rather than the offshore staff, that must acquire knowledge in order to make the offshore engagement work more effectively. Ideally, the client elevates its internal software practices to CMM Level 3 or ISO 9001 before beginning the offshore work. However, this is uncommon because moving up the CMM levels requires, at the very least, 6 months per level. Most client organizations are not interested in making such a commitment and instead, a sufficient alternative is to faithfully comply with the requirements of the provider's CMM processes.

The last two types of knowledge are largely *tacit* knowledge that cannot be easily transferred via training and documentation. The first of these knowledge transfer types, *domain* knowledge encompasses specialized business, scientific, algorithmic, or artistic knowledge.

Some organizations recognize the difficulty of domain knowledge transfer and are proactive in managing it. For example, when the internal information systems staff at Wal-Mart, the largest retailer in the world, embarks on a project, they go through a rotation in the end-user area. If, for example, they are tasked with working on the Point-of-Sales systems, they go work at the registers for a few weeks helping customers to checkout their purchases.

Domain knowledge transfer often fails when the offshore engineers do not fully understand how the software will be used. Domain knowledge is vital in nearly all activities offshored. When application maintenance tasks are transferred offshore, then the new software engineers need to understand the implicit, tacit link between the code, the data, the business rules, and the ways they are used. When new software development is conducted offshore, knowledge transfer is straightforward if the software can be *precisely* specified but software is rarely specified precisely. Sometimes the difficulty revolves around usability issues, such as how credit cards are used or how decision-making is actually conducted. The case study in this chapter (called "Eating your own dog food") describes such difficulties.

The last knowledge transfer type is labeled "Work and cultural norms." This is the most difficult type of knowledge transfer. It encompasses deeply set cultural norms that have to do with foreign cultures, how foreign clients work, as well as organizational cultures. The two chapters that follow (Chapters 8 and 9) cover these issues in greater depth.

Since tacit knowledge transfer is primarily about transferring knowledge into the minds of people (rather than into systems), the principle solution revolves around face-to-face interactions. Human beings traveling between distant sites are the principle conduit for knowledge transfer. Travel can be of any duration, but extended rotations may be necessary.

One large American financial services company institutionalized knowledge transfer in an interesting way. The firm had been working offshore for three years and regularly rotated offshore staff to the USA every 3 months. Four to eight people would come in from India every 3 months. Once they returned then another group came to the USA. This would be repeated every 3 months. Over time, some of the offshore staff had been to the USA several times. Naturally, this was costly, reducing the offshore savings, but the company felt that the expense was justified because of successful knowledge transfer.

Offshore providers have recognized that knowledge transfer is a serious difficulty and needs to be managed properly. One offshore provider developed a knowledge transfer methodology it called knowledge acquisition process (KAP).[5] Such knowledge transfer methodologies are based on two principles: extensive documentation and intentional face-to-face interaction. Specifically, they call for onsite presence in the first few months, shadowing employees to learn their jobs, and documenting the knowledge as much as possible into service level agreements (SLA), plans, and technical documentation.

Case study Knowledge transfer by "Eating Your Own Dog Food"

One successful approach to knowledge transfer is to use a method with a colorful expression: "eating your own dog food." The term gained wide usage after it was popularized by Microsoft in the early 1990s. In this method, once the product is minimally usable, the programmers use the software product on a day-to-day basis, to synchronize their activities, or as their own development platform. By using the software product the programmers are able to make better minute design decisions and are quicker to fix its bugs.

GroupSystems.com, a small US-based software company, decided to outsource offshore all software development activities for a major release of its flagship product – a platform that supports collaborative decision-making. In 2001 it was time to create a new vision for the product and change its underlying architecture. This was a significant undertaking for this small firm. GroupSystems contracted with IISC, one of the top five Indian offshore providers (IISC is an alias).

The new product included a great deal of tacit, "domain" knowledge, as Bob, the GroupSystems R&D Director, emphasized:

> *"Relative to most software products, the user interface requirements for such a collaboration tool are more nuanced and sensitive to subtle variation."*

Problems in transferring knowledge began to surface soon after development began. While the offshore programmers were skilled at coding, they did not grasp the product vision. Technical specifications could not capture all the usability possibilities in full. For example, Bob had written 37 pages of high-level requirements on just *one* feature of the user interface. This one feature included subtle usability capabilities: how to manage selection of shared text blocks on one screen while other users (on other screens) were simultaneously adding to, modifying, moving, and deleting text in the same document. Even with such a level of detail, however, he acknowledged that he was unable to anticipate and specify every small design detail in the user interface.

The written specifications and frequent trips across the Pacific could do only so much. None of the 25 offshore programmers had any experience using, or developing, a collaborative product and did not fully absorb when and how it should be used. Furthermore, the programmers were reluctant to ask clarification questions. Since they did not ask, they were forced to make small, seemingly arbitrary design choices; these are choices which often thwarted the product's intended uses. To make matters worse, on numerous occasions the offshore programmers resisted design choices made by the GroupSystems staff, arguing that the specifications were unnecessarily fussy, and that "there are faster, easier ways to do it."

Thirteen months slipped by and the product fell further and further behind. More than 250,000 lines of code were written that eventually had to be thrown away. The project was at a crossroads and the future of GroupSystems was in doubt. Managers from IISC and GroupSystems held a crisis summit. As a result, IISC re-organized the project team, and assigned Arun, a seasoned Project Manager, to this troubled project.

The new project team, though, had a typical composition: 40 very young, bright software engineers. Most were just a few years out of the local second-tier university. All had engineering or computer science degrees, but none had any training in business. Nor were they ever involved in the subtle issues of collaborative decision-making in an organization.

Arun drove his team hard. Most importantly, he taught his programmers how to ask questions and then how to listen. He taught them to seek clarification of any ambiguity or doubt before they wrote code. The team held trans-continental phone conferences on a daily basis to review progress and clarify concepts.

It was at this point that GroupSystems hit on an idea. Since the programmers were still struggling to understand the product, then why not have the team use the very product that they were developing? In other words, why not have them eat their own dog food?

As soon as the new code was barely stable, Bob traveled to India and spent a day creating a rudimentary simulation exercise for the programming team. He created a fictional scenario in which the IISC team would "meet" using the GroupSystems product. The goal of this meeting would be to write a proposal in response to a request for proposal (RFP) for a payroll system that will be bid on by IISC. A mock script was created for each step, complete with an embedded IISC logo in each module to make it look as realistic as possible. Thirty team members were then distributed across the IISC campus and began a distributed meeting using the software product. The team was walked through five typical steps in proposal writing, moving through a complete proposal-creation cycle.

"It was then that I saw all the light bulbs switch on," recalled Bob. "Not only was I deluged with good questions and comments, but I got comments along the lines of: 'Now, I understand why you asked for the drag and drop design the way you did.'"

From that point onward, the Indian programmers used the product several times a week, to see how it felt and how it performed when used as it would be in the field. The usability problems that had plagued the project now melted away. The quality of delivered code soared was compared to industry benchmarks. A full commercial product was released 12 months later.

Change Management

Organizational Change Management: Making changes in a planned, systematic fashion with the focus on instilling new attitudes, behaviors, and consensus building within an organization.

Employees and managers around the world have offered many valid objections to offshoring. Here is a smattering: it's not secure (our data will be compromised); the savings are overstated (here is a magazine article about an offshore fiasco); our employees will be laid off; what happens if our systems go down because of offshore instability? (there was a terrorist bombing there yesterday); "it's core, so we'll never offshore" (my system is too important to the firm); there are too many miscommunication problems (they don't speak English very well). Or, one unspoken objection: "I don't really want to deal with people from that country." This is merely a partial list.

In short, offshoring, which represents an organizational change, meets resistance inside organizations. Such is the case of a major US corporation described in the next case study later in this chapter.

In order to address such resistance, *Change Management* approaches include: implementing measures and reward systems to motivate offshoring, creating new organizational structures to support change, internal selling, funding demonstration projects, education about offshoring, and implementing human resource policies to reassure employees of their positions. We cover each of these approaches below.

Change can be hastened by organizational measures and rewards linked to these measures. Such measures have been used successfully for offshoring in quite a number of American companies. For example, a specific goal can be set to increase offshore headcount from 5% to 8% within 18 months. GE's famous 70:70:70 mandate for offshoring was described in Chapter 5 and may be the first of these offshore measures. Once the offshore goal is established, then executives need to reward and enforce meeting this goal.

Change is also brought about by creating the right organizational structures. Offshoring is often managed by centralized units within large corporations, sometimes from within the offices of the Chief Information Officer (CIO) or the Chief Technology Officer (CTO). We have found that many of these units have adopted the word "global" in their names, and thus: Global Engineering, Global Resources, Global Services. *Global* is a word that creates less resistance than *offshore*. We will refer to this generic organizational unit as the Global Sourcing Unit. The Global Sourcing Unit may be a part of a Global Information Office (see the governance section later in this chapter).

The role of the Global Sourcing Unit is to assist various internal users in assessing and migrating offshore. It houses offshore knowledge bases, such as provider and country information. In some instances, the Global Sourcing Unit implements the new measurement and reward systems, mentioned above, that encourage project-level decision-makers to find the best software resources, inside or outside the corporation. It may also work with the various corporate divisions in conducting an inventory of the corporate systems in order to identify the best candidates for offshoring. In so doing it should also help identify the roughly 20% of IT functions that are truly too complex or proprietary to be considered for offshoring.

The Global Sourcing Unit is sometimes headed by the "offshore champion," who is usually a seasoned manager. The offshore champion is the change agent, and plays the typical evangelist role. She becomes a "salesperson," generating excitement for the cause and "selling" the offshore vision internally. She communicates the offshore vision and strategy to reluctant managers. She sets up seminars and develops internal sales brochures. Like politicians, the offshore champions do not see any drawbacks in their cause: it is clearly the way to go. Get on the train! The offshore champion needs to be well respected. In fact, the most effective change management path is to earn respect, rather than force compliance.

The Global Sourcing Unit is also the catalyst for demonstration projects to bring about change. One of the quickest offshore change management programs took place at a large American company that budgeted several million dollars for such a catalyst program. The offshore champions roamed around the organization and dangled money in front of various IT managers saying, in effect: "We'll give you extra budget to do this project if you offshore it." Within less than a year the company completed 80 (!) offshore demonstration projects and had many offshore converts.

A different approach to change is via internal education about offshoring. Seminars on offshoring should be informative, motivating, and help prepare for the change. Such educational opportunities can be targeted at both business managers and IT professionals.

Of course, one of the greatest obstacles to offshoring is the fear that one's job will be replaced. Americans call this "restructuring" and the British call this "redundancies." While some organizations have chosen to lay off many employees at once, others have chosen a slower route, by reducing onshore headcount through attrition. In order to reduce its employees' offshore anxiety, IBM announced publicly that it would set up a 25 million USD training fund to retrain employees in the USA and UK whose jobs were threatened by offshoring.[6]

Yet another backlash-related issue is whether to be open and public about offshoring. Managers weigh the trade-offs of keeping offshoring plans secret or communicating them openly to demonstrate honesty to their employees and their communities.

Case study The ups and downs of building support for offshoring at a giant US corporation

The case of General Marvel illustrates the prolonged journey of offshoring acceptance within large organizations.

General Marvel is one of the 100 largest corporations in the USA. The case of General Marvel is an actual case, but all identifying details have been disguised.

General Marvel's IT department has one of the longest offshore histories, having been involved in offshoring since 1990. Nevertheless, 14 years after beginning IT offshoring General Marvel is still slowly overcoming resistance.

General Marvel's offshore saga began entirely by chance. In 1990 one of its software product providers, PrexiSoft, had developed its product in India. PrexiSoft convinced General Marvel to try outsourcing a project to India. "We assumed that since they were experts at this, that with their help we too would be successful," recalls one manager. The project was a disaster and it had to be brought back in-house and redone.

In 1993 General Marvel was growing quickly but about to hit a wall. Its all important, corporate-wide product codes were only seven digits wide and the company was about to run out of possible codes. This became the "Overhaul" project. General Marvel's IT managers began shopping for providers to solve this problem. They decided to divide the work among two US providers and one Indian provider.

Once the work was under way, General Marvel assessed the results and found that the Indian provider performed this tedious work at higher quality levels and at lower cost. The two US providers were phased out. At its peak, Overhaul employed 100 offshore programmers in the project.

In 1994, with the Overhaul project completed successfully, General Marvel's IT managers had the foresight to embark on Y2K remediation. The work was offshored to India and completed successfully within a year. General Marvel had become one of the first major companies in the world to successfully complete its Y2K remediation!

Thus, by 1995, General Marvel, in spite of its early stumble, had two offshore successes. It had internal champions in its IT division who had developed relationships with providers and with provider managers. Some of General Marvel's managers had already been to India. The offshore successes led to recognition from top management: the successful managers on these projects were promoted. "I like the offshore model" one manager recalls stating to his colleagues.

The principal offshore provider, RMI, had grown to know General Marvel's systems from the inside. RMI proposed that it begin to take over system management and maintenance. Now, for the first time, the offshore provider was becoming entrenched, moving into long-term systems support activities. Up until then, General Marvel was offshoring projects, such as Y2K remediation, that had an ending date and that no one at General Marvel's IT department really wanted to do.

General Marvel's IT Groups were composed of internal programming crews and contractors (non-employees with individual contracts) who had worked on company systems for years. Each crew was fiercely loyal to its members and to the systems they serviced. As RMI began to take ownership of some IT systems, individual contractors were terminated and the long-standing crews began to get nervous. Resistance to offshoring had begun.

General Marvel had three major IT Groups corresponding to its main business functions: Production, Distribution, and Finance. The Production IT Group was managed by Tommy, an old-timer at the corporation. Tommy became an offshore proponent looking for opportunities to offshore whenever he could. In 2001 he promoted Tandy Danielson, then a young and ambitious software team leader, to manage an offshore project. First she managed a small Time & Material project with the Indian provider. Then, promoted to Project Manager, Tandy managed a $5 million Fixed Price project with most personnel in India. Both projects were "massive successes." Tandy demonstrated that it was possible to manage critical projects in which most of the developers were in India. Tandy became a "hero," receiving the President's Award for Outstanding Junior Manager. She also became an offshore proponent, though this was short-lived. She was moved out of the Production IT Group to the Distribution IT Group, which was largely resistant to offshoring, and her work on offshore projects stopped.

Nevertheless, RMI and other offshore providers were performing more work on General Marvel's legacy IT systems. But the dramatic cost savings were not materializing. The IT budget was still growing out of control, and IT headcount was growing.

Some IT managers were not comfortable letting go. They were more comfortable asking the offshore providers to supply them with onsite personnel. The providers obliged. Indian personnel were arriving at the local airport with their families, teaming with General Marvel's staff and working on projects. Quite a few of them were settling down, buying homes in the city. Meanwhile, the company's overall costs, instead of declining, were increasing. IT managers were not being pushed to increase the ratio of offshore personnel, so they did what was comfortable and familiar, they kept them onsite. "I cannot get rid of Singh and Aggarwal," said one IT manager, "they are critical to my system team." And so they stayed and stayed and stayed.

Maryann was one such IT manager. She spent most of her professional career at General Marvel and was a well-respected IT manager with strong technical abilities. One of Maryann's strengths was that she was always very involved in carefully screening individuals to work on her crew. She had a good track record at completing projects with her hand-picked software engineers. She applied this ethos to her relationship with RMI. She carefully screened each of the Indian software engineers and preferred that as many of them as possible be onsite. "I think that for her the provider relationship is irrelevant" said one of her colleagues, "she sees the offshore provider as a supplier of skilled individuals, instead of trusting that the provider will get the job done."

Speaking about Maryann and other IT managers who were not behind offshoring, one of the offshore advocates said, "You hear lots of excuses from them about offshore failures and the difficulties involved, but it takes hard work and then it succeeds. You write good requirements, freeze them, and then overlay good project management. It is not easy, but it can be done. It's a learning curve."

He added, "It is true that the Production IT Group has always been the leader in offshoring because their systems were more amenable to the offshore model. The Distribution IT Group deals mostly with rolling out systems for the distributors, while the Finance IT Group does mostly packages. Both are less amenable to offshore."

In 2002 Mike Hudson, the new CIO, became irritated with this situation, saying in a meeting:

> *"Why are we offshoring when most of their people (the providers') are here working onsite!"*

Driven by this and other factors, Hudson decided on a new change strategy: moving to centralization, standardization, and shared services. The multiple offshore provider approach that had been in existence by default was eliminated in favor of one primary Indian provider, one of the Tier-1 Indian providers. IT application support became a centralized function. "If you have to fight the (offshore) battle project by project, it is much harder. With centralized management control it becomes easier," argued one manager.

When asked about soft persuasion, a manager replied in frustration, "we've been using soft persuasion. We've been using it for 10 years." Hudson tried a "stick approach" by setting some offshore mandates. But, when the mandates were not met, he did not hold the Group IT managers accountable for failing these mandates. "People made excuses," said one source.

The change was coming from local champions. In 2003, one of General Marvel's three major IT Groups, the Finance IT Group, did the least offshoring. This changed when Ted Chung became the new Finance IT Group head. Chung was an offshore champion. And, within 12 months, the Finance IT Group became the most active offshore user.

By 2004 General Marvel's IT workforce stood at 3000 including outsiders. Of this workforce, 15% were offshore in India at lower costs and another 15% were onsite in the USA, but were employees of an Indian-based provider.

After almost 15 years of offshoring, one manager quipped: "we're just getting started to change people's minds."

Concluding observations

- General Marvel used many of the ingredients of the change management recipe: it implemented new organizational structures, it had successful demonstration projects, it instituted rewards, and it fostered offshore leaders. However, none of these ingredients made dramatic changes in General Marvel's organizational practices or culture. None seemed to create an offshore momentum.
- All the same, the gradual, small changes that occurred over many years may well be a mark of success. While there was no offshore zeal, there was considerable acceptance. Furthermore, there were no episodes of broad employee backlash to offshoring. In late 2004 General Marvel announced that it would open its own center in India in 2005.

Most managers who interview in the business press tend to favor the latter approach for obvious reasons.

Governance in offshore outsourcing

Short definitions for the notion of governance are not terribly useful, so a detailed one is given here that will be expanded upon in the rest of this section. Governance deals with aligning the strategies and goals of the client organization with the provider, cascading these goals down the respective organizational hierarchies, creating appropriate organization structures for both client and provider to achieve these goals, creating relationships and open communication channels, creating a control and monitoring framework, and measuring the provider performance. In sum, governance is a joint responsibility of the client and the provider in which both parties set up and agree to roles, responsibilities, relationships, measures, problem solving processes (escalation), scope of work, and termination processes.[7]

Three principles permeate the notion of governance. First, there need to be many communications channels between the client and the provider that are open at all times for effective dialogue. Second is *relationship management*, in which there is mutual recognition of interests, as well as trust between individuals. Third, there needs to be constant reporting by the provider at all levels (operational, tactical, business), and at various appropriate intervals: whether this be by the minute, the day, or the month.

Governance is an expensive part of the overhead needed in offshore outsourcing, estimated at 5% of the outsourced contract value for domestic outsourcing and 6–7% for offshore outsourcing. Due to high costs, governance structures are only instituted above contract amounts of 50,000–100,000 euros and typically in engagements between large global firms and large providers.

The higher governance costs stem from several factors. Global firms typically operate in multiple geographical regions and therefore this leads to greater coordination effort in software development projects. The geographical spread also has legal implications since multiple national laws apply to infrastructure management engagements. There are higher personnel costs since the client has to designate governance roles at various locations and with specialized expertise. For example, international business experience is desirable: a US-based client working with a Brazil-based provider should try to find a US-based contract manager with some Brazilian experience to manage the offshore provider. Lastly, the global firm usually suffers from multiple standards and multiple hardware and software platforms across the organization.

The Service Level Agreement (SLA)

The SLA is a central element of governance. The SLA defines the contracted quality and quantity of the services along a number of dimensions. It is the operational and legal mechanism by which the client claims the contracted services from the provider.

Before we delve further into the SLA, we reiterate the two types of IT services that can be offshored: software development and infrastructure management. Briefly, software development refers to early life-cycle activities including design and coding. Infrastructure management includes operational services, helpdesk, and performance trouble-shooting. The key difference between the two revolves around duration and interaction. Software development is a relatively short-term engagement with ever-changing requirements that involves frequent interaction with end users. Therefore the key difficulties are in collaboration issues such as communication across cultures. On the other hand, infrastructure management is a long-term engagement, with relatively limited interaction with end users; it is process driven, with fairly stable requirements.

The focus in an SLA for software development is on completion of deliverables on time and according to the contracted requirements. For example: "The application has to be implemented by December 31, 2005 according to the specification in Appendix A. The implementation has to be executed according to the implementation plan in Appendix B including testing and approval of the client." In contrast, an SLA for system management focuses on service hours, availability, and downtime (see Table 7.1).

Of the two types of IT services, software development and infrastructure management, discontinuity risks are relatively low for the first. For example, downtime of 1–2 days is usually acceptable. For infrastructure management, on the other hand, risks are high, since they involve managing critical client systems at the heart of the client's business. This means that there are high threshold requirements for the availability of communication facilities. In cases of instability or crisis offshore, providers should be able to transfer service provisioning to another data center in another country in order to meet the agreed service levels.

As part of the SLA, larger offshore outsourcing relationships often use a Balanced Scorecard for reporting and discussion purposes. This is a kind of dashboard for the client to monitor the provider's performance. The indicators on a Balanced Scorecard are more business-oriented than technical or operational. Scorecards typically have four perspectives: business processes (e.g., lower personnel costs by 40%, decrease inventory costs by 15%), customer perspective (e.g., ability to add new product offering within 4 weeks), organizational learning (e.g., all employees are able to use the implemented IT system prior to the end of the year), and financial perspective (e.g., increase sales by 10%). Unlike the service levels that appear in Table 7.1, the scorecard aligns the business goals of the customer more closely with the provider's. Scorecards may be combined with performance incentives for the provider, though this has not yet become a common practice in offshoring.

A common SLA component is the specification of methods to improve delivery and reduce risks. The customer and provider choose from the "alphabet soup" of methods for this. For software development they include: CMM, ISO 9001, and PMBOK. For support they include: ITIL, Six Sigma, and BS 7799 security certification.

Governance structures and key roles

In this section we examine a generic offshore governance structure and then compare it with an example from a global corporation to see how this structure is adapted, and why.

A proper generic governance structure appears in the diagram of Figure 7.2 along with a relationship and responsibilities matrix in Table 7.2. There are several governance principles that drive these displays. First, the roles and responsibilities in these large

Table 7.1 Example of SLA for support of enterprise systems services, such as SAP

Service level elements	Service levels	Penalty	Measurement and reporting period

Service hours
• System type 1
• System type 2
	• 7*24 hours
	• Europe: M–F
	  07:30–16:30 GMT;
	  USA: M–F
	  14:30–23:30 GMT
		• Not applicable
		• Not applicable
			None

Availability
• System type 1
 	
• System type 2
	• 99.8%, based on a rolling
	  3-month average
	• 99.0%, based on a rolling
	  6-month average
		• 25% of monthly
		  cost
		• 25% of monthly
		  cost
			Monthly
			
			Monthly

Maximum downtime,
cumulative per year
inside service window
	16 hours (for System type 1
	servers), (limited to
	production servers)
		• 50,000 USD
		  for each
		  additional hour
			Monthly

Maximum number of unscheduled downs >4 hours during the last 12 months • System type 1 • System type 2	 • 2 • 4	 • Not applicable • Not applicable	Bi-monthly
Maximum number of unscheduled downs during the last 12 months • System type 1 • System type 2	 • 6 • 12	 • Not applicable • Not applicable	Bi-monthly

Response time (as
reported by the
standard internal
SAP tool)
	• 90% within 1 second,
	  (limited to
	  production systems)
	• 95% within 4 seconds
	  (limited to production
	  systems)
		• 25% of monthly
		  cost
		
		• 25% of monthly
		  cost
			Bi-monthly

organizations are set up in a symmetrical structure in terms of geographical location and hierarchy. Second, multiple channels of communication are established and aligned. Third, cascading levels of strategy, partnership, relationship, coordination, and minute-to-minute service delivery are carefully defined between the two sides. Fourth, is the proximity principle. Governance structures in offshore outsourcing benefit from

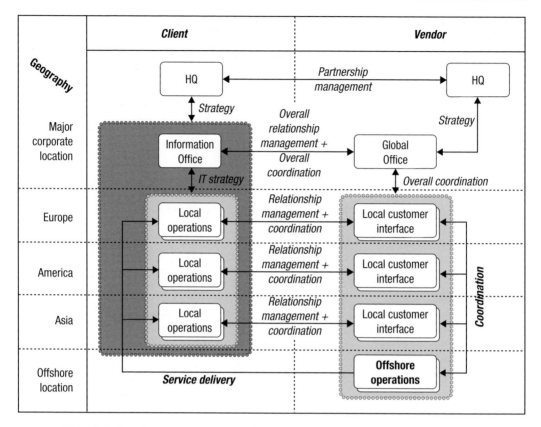

Figure 7.2 Generic governance structure between client and provider.

proximity. Technically there is no need to have a local customer interface but communication is improved due to proximity.

The two most important governance units are the *Information Office* (in case of software development sometimes called Project Management Office (PMO)) on the client side and the *Global Office* on the provider side. The Information Office and the Global Office are the highest-level units on both sides: all disagreements escalate eventually to these two offices, and on rare occasions they escalate further to the executive levels of each side. The client's Information Office manages both local and offshore outsourcing and oversees coordination of operations in different countries.

Some large firms have not established a robust Information Office and it is therefore the provider's responsibility to help them structure and build such a unit. It is in the provider's interest to deal with a strong Information Office because this unit tends to structure and formalize the business requirements. The provider is also acting in its self-interest because such a unit will enhance the relationship.

In offshore outsourcing, some clients establish a Single Point of Contact. This does not necessarily contribute to effective governance. Effective offshore governance requires multiple channels of communication at different levels and different locations. These

Table 7.2 Roles and responsibilities within governance structures

| Responsibilities | Client | | | Provider | | | |
	HQ client	Information Office	Local operations	HQ provider	Global Office	Local customer interface	Offshore operations
IT strategy		CIO					
Partnership management	Board member responsible for IT			CEO	Global Sales Executive		
Relationship management		CIO + Information Managers	Business Unit Manager		Global Customer Unit Manager	Sales Manager	
Coordination		Information Managers + Service Delivery Supervisor	Business Unit Manager		Global Customer Unit Manager + Contract Manager	Local Customer Unit Manager	Service Delivery Manager
Service delivery			Business Unit Manager + End users		Process experts	The IT profes- sionals	Service Delivery Manager + IT professionals

channels are needed at the strategic relationship level (the Information Office), at the tactical level (IT and business management of the local operations), and at the service delivery interface on the operational level (IT management of the local operations).

The client manages the provider via two key documents introduced earlier: the SLA (at the operational and tactical level) and the Balanced Scorecard (at the strategic level). The service reports provide evidence of the provider's performance relative to the SLA. Some clients manage their multiple provider contracts as a portfolio, in order to avoid a provider lock-in, and threaten to replace non-performing providers based on sub-par reporting vis-à-vis the SLA.

On the provider side, the Global Office is responsible for overall coordination of all aspects of the client's services, wherever in the world they may be. Importantly, the Global Office is situated near (or if possible at) the client home office.[8]

Let's take a look an actual offshore governance structure and see how it differs from the generic structure and why. Figure 7.3 shows the governance structure between the large France-headquartered provider Atos Origin and one of its large global clients that we will call CPG Inc.

CPG Inc. is an actual US-based consumer packaged goods company with extensive European operations and a European Information Office in Belgium. By 2002, CPG Inc. had been buying domestic outsourcing services from Atos Origin for over 10 years. Initially Atos Origin was offering these business-critical IT services out of its Belgium delivery center directly to CPG Inc. operating companies all over Europe and Asia.

In 2002 CPG Inc. decided to offshore some of these operations in order to reduce costs. The services chosen to offshore in this case were infrastructure management and included activities such as operations desk, routine changes for DNS management, routine changes to printers, patching midrange systems, file system management, server reboots, and kernel parameter changes. The annual amount of the contract was about 100,000 euros, representing a small percent of the contract value of the Atos Origin CPG Inc. relationship. The provider chose to deliver the services in a combination format from both Poland and India despite the additional communication and coordination costs. This need was acute at the contract signing-time because of the heightened tensions between India and Pakistan. If staff in both locations were actively involved in daily operations they could easily take over for one another in case of crisis.

The three arrows in Figure 7.3 point to three areas of governance structure where the CPG Inc.–Atos Origin case differs from the generic governance structures pictured in Figure 7.2.

The first arrow points to a special structure in the provider's Global Office. Since CPG Inc. is a US-based firm it was more effective to have the provider's global client executive based in the USA rather than at the European Global Office which coordinated the European service provision. But this resulted in a split responsibility for the overall relationship: the Belgium-based Customer Unit Manager was limited to the offshore relationship, while the global client executive was responsible for the entire relationship. This split responsibility required frequent (twice a week) conference calls between the two.

The second arrow points to the key roles created by Atos Origin. The provider had to implement an alignment between the Polish and Indian service delivery groups so that in case of emergency the service provisioning could be transferred to the other location. Another alignment was the shift work involved in 24*7 operations. The two locations split the shift work since the cost of the night shift in India was significantly higher than the day shift. Atos Origin assigned 11 full-time equivalents (FTEs) to the contract: 7 in Poland, 4 in India, and a half-time Poland-based Service Delivery Manager. The Polish Service Delivery Manager acted as the interface to all the client's local operations in both Europe and Asia. Most of his attention was devoted to coordinating the IT professionals in Poland and India, since there were large cultural differences between the two.

The third arrow points to the two additional governance roles that the client, CPG Inc., needed to create at its main European center: a Program Manager and a Contract Manager. These two roles represent the main increase in governance costs relative

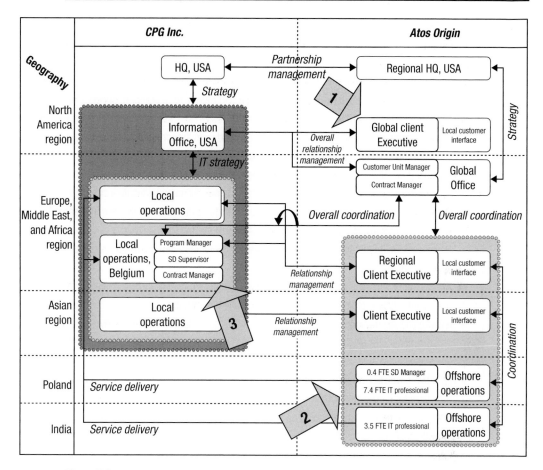

Figure 7.3 Governance structure of the CPG Inc. offshore outsourcing of infrastructure management.

to the previous domestic outsourcing. The client anticipated that the new offshore outsourcing engagement would require more staff time since CPG Inc. in Belgium had limited experience in offshoring systems management relationships. Hence, these governance roles were understood to be somewhat temporary: as the client gained offshoring experience the governance staffing needs would be reduced. There were also other considerations for these roles. First, the assigned client Program Manager had a long history with Atos Origin and was therefore needed for his experience in oversight. Second, the Contract Manager was to gather the Balanced Scorecard data that came from local operations in Belgium.

Concluding lessons

- Identify those offshore projects that require complex knowledge transfer and manage these engagements patiently, devoting sufficient time and money to make them successful.
- Successful knowledge transfer revolves around two essentials: staff rotations and, separately, careful knowledge documentation. There are no short-cuts around these two essentials.
- Organizational acceptance of offshoring often requires creating new centralized organizational units, such as "global sourcing units."
- Organizational acceptance of offshoring requires internal selling and education efforts.
- Organizational acceptance of offshoring requires organizational goals that are implemented through measurement and rewards for managers.
- Organizational acceptance of offshoring is helped by funding demonstration projects out of special budgets.
- Organizational acceptance of offshoring is most successful when it is spearheaded by a respected offshore champion that acts as a catalyst for change.
- As with change management, offshore outsourcing governance requires a new organizational unit, such as an Information Office, that centralizes many of the coordination and relationship functions.
- Effective governance requires many open channels of communication between client and provider at multiple hierarchy levels.
- Effective governance is fostered by good personal relations and trust between the individuals in the liaison roles.
- Effective governance can benefit from specific documents: the SLA and, if appropriate, a Balanced Scorecard.
- Effective governance requires personnel assigned to new roles. The governance overhead that needs to be budgeted is about 6–7% of the contract value.

8 Overcoming distance and time

All things being equal, any manager would prefer to manage a co-located team rather than a distributed team.

Offshoring requires *distributed collaboration* in which people work across distance and time. This is a key difficulty in offshoring. In this chapter, we first take a close look at the root of the problem: Why is it so difficult to collaborate across distance and time? We then present the many small solutions to this problem. The difficulties in distributed collaboration cannot be eliminated, but they can be mitigated somewhat through a mosaic of solutions described in this chapter: applying principles of formalisms and informalisms, managing time differences, using a mix of collaborative technologies, selecting the right staff, and designing the optimal organizational structure.

We like to be close

In spite of the hype about our new "virtual world" in which "distance is dead," we humans like proximity. We perform better when we are close together. We thrive when we have face-to-face interaction.[1] We crave proximity.[2]

In order to understand why distributed work is more difficult for us humans, Kiesler and Cummings[3] gathered the results of decades of group psychology research. We begin by summarizing their thought-provoking findings.

The mere proximity to another human introduces a "social facilitation effect." That is, our physiologic performance changes: alertness increases, our heart rate goes up, and our blood pressure increases. Television producers introduced laugh tracks to comedy shows because we all tend to laugh when others laugh; we smile when others smile. Researchers found that when we experience an event with someone else the event tends to be more memorable. Even more interesting, food tastes better when we are with others. We feel more involved when we are with others; we are "energized." Simply being familiar with someone tends to increase liking and to heighten identity to the group or team. Simply being closer to someone else makes us more likely to conform, or obey orders. In one classic 1970s experiment, subjects were more likely to apply dangerous electrical shocks to their fellow students when the orders were received from someone close by.

As social beings, we behave differently in different settings (in a bar, a church, or a school). We all work in such a *shared social setting*, such as an office. We get territorial about social setting and about our personal territory. We all have territorial "bubbles" around us. We don't like it when others puncture these bubbles. We have territories around our desk space or the group of team cubicles. The territory tends to strengthen the ties we develop with the group of people we work with. Furthermore, this shared social setting leads us to be more satisfied with that team at large. Thus, cohesive teams tend to sit together. The problem with our territorial attachment is that it interferes with our identification with the larger distributed team: the distant programmer in Manila is not a member of *our* territory.

Proximity also leads to spontaneous communication. These are chance-encounters that lead to conversation: an encounter in the hallway, in the office kitchen area, or before a meeting. Americans call these spontaneous conversations "water cooler" conversations, still using an image that is fast disappearing from most American buildings, namely a dispenser of cooled, piped water. These spontaneous conversations are enormously powerful in organizational life because it is through these chance-encounters that we get to know what others are working on and how well they are progressing. We do quick problem solving, thus facilitating coordination. These chance conversations are also important to non-task objectives: they help solidify relationships. We are more likely to like people we chat with and we are more likely to be influenced by these people.

Office distance has the highest impact on spontaneous communication. Many of us have discovered from experience that we hardly ever see Ariel, who has an office just one building away from us. Tom Allen was the first to confirm this when he measured workers' chance-encounters and found that spontaneous encounters for people whose offices were more than 30 m apart reduced to a chance of only 10% per day (Figure 8.1).

Silicon Valley illustrates how proximity makes a difference. Most high-tech firms benefit from being in high-tech *clusters*, the most successful of which is Silicon Valley.

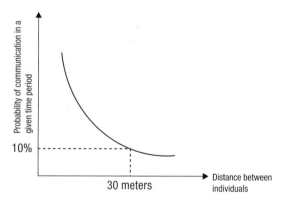

Figure 8.1 The Allen curve showing the probability of spontaneous communication between co-workers.[4]

All else being equal, each firm in the cluster is better off from its location in the cluster rather than in a distant, remote location. Innovative firms benefit from the many face-to-face interactions that clusters enable.

Understanding the problems of distance

We use a physics metaphor to frame the problem of distance.[5] A *centrifugal* force is a physical force that propels an object away from the center. Distributed software collaboration is like a centrifugal force that propels the team members apart from each other (see Figure 8.2). Each of the five centrifugal forces is introduced and explained here.

Communication breakdown

We human beings communicate best when we are close. Why? Because we are conditioned to convey and read each other via more than the naked text that we utter. The way the text is delivered, via tone of voice, the pauses in our speech, an accent that we place on a phrase, all convey so much. Furthermore, our body tells a story. We open our eyes, furrow our brow, smile, frown, gesture with our hands, point with our fingers. All of these are part of our communication that is lost when we try to send a message, or convey a vision, over a narrow communication channel, such as e-mail. Some say that 80% of the message we convey is in the non-naked text. In fact, some go further and make an evolutionary argument: during most of our evolution as a species, our ancestors communicated primarily face-to-face, so our brains are hard-wired for this form of communication.[6]

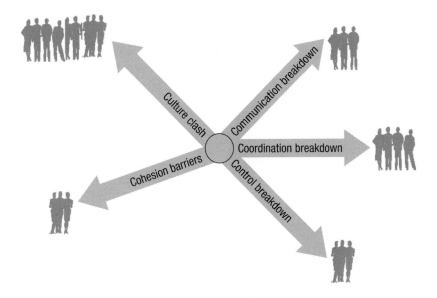

Figure 8.2 The five centrifugal forces that make distributed software work difficult.

We also know that the more complex and important the task, the more we need face-to-face communication, or at least a telephone call.[7] A serious design session is difficult to do over distance. A large contract negotiation is rarely done over distance. Requirements gathering will not be comprehensive when done over distance.

Successful communication has an exacting definition that is worth pondering: communication is complete when the information has been transmitted, received, acknowledged, understood, and acted upon. Accordingly, when our messages are not understood properly, which is much more likely when we are far apart, we say that we have a *miscommunication*. In distributed collaboration this miscommunication leads to delays (because of the need to clarify); to rework (because we didn't really understand what was meant); and, most painfully, miscommunication leads to conflict because, all of us, in all countries, get fussy over personal slights that we interpret in the message text.

Coordination breakdown

Software development cycles require frequent "small adjustments."[8] Coordination is the act of making those adjustments, or more formally, it is the act of integrating each task and organizational unit, so that it contributes to the overall objective. We coordinate work via countless small adjustments: a question, a request for clarification, a small improvement, an *ad hoc* solution resulting from a one-minute chat while standing in line at the cafe. When you are working in the same location, you are *aware* of who to contact for help because she's one floor below you and she sits next to Josepha with whom you play tennis. In fact, awareness and its close cousin "shared knowledge" are vital elements of coordination.[9]

In distributed collaboration all of these small adjustments are difficult, since much of our coordination results from spontaneous conversations or from the small cues about what is happening in the project. And, time separation makes all this worse. When coordination slows or breaks down, several dynamics occur. Problem solving gets delayed again and again until it becomes very expensive to fix. Things move along the wrong path for so long that it becomes difficult to renegotiate all the sides back on track.

Control breakdown

Control is the process of adhering to goals, policies, and standards. Years of experience teach us that successful control takes place when managers can roam around to see, observe, and dialogue with their staff. Hence, MBWA (Management By Walking Around).

When a software project manager is supervising developers many kilometers away, roaming around and getting a "feel" for what's happening becomes an unusual event. Sometimes it never happens at all. And, when managers cannot roam, they have to rely on collecting information and imposing their will by means of technology, through telephone and e-mail. This is less effective then face-to-face. As an added insult, managers tend to pay more attention to those who are close to them and less to those far away, and thus the saying: "out of sight, out of mind." The result of poor control in distributed collaboration is wasteful duplication of effort, discovering problems late, and the subsequent need for rework.

Cohesion barriers

Groups that are close together jell and bond.[10] People get to like each other, trust each other, help each other, work harder for each other. It is not surprising that in distributed collaboration it is very difficult to foster cohesion unless there is *history*. That is, unless the people have worked together before in the same location.

Distributed teams are more diverse, by culture and organizational background. Such teams are less likely to rally around the same vision. Often, members are working on multiple projects and on multiple teams. For example, at Intel, one of the most dispersed technology companies, a 2003 survey found that 63% of Intel employees were members of at least two teams and most of these teams were, themselves, virtual teams. (The Intel case appears later in this chapter.)

Cohesive project teams trust each other. But building this trust takes time, even for co-located teams, let alone for distributed teams. Trust presents a paradox for distributed collaboration: on the one hand you need trust in order to work effectively together over distance; on the other hand it is quite difficult to develop trust over distance.

Trust is an important, but elusive, concept that merits a brief digression. First, it should be evident that the opposite of trust, known as *mistrust*, is to be avoided. Mistrust, once it begins, can ooze and fester and destroy everything that it touches. So, it is critical to actively reduce the chances that mistrust appears.

Our favorite definition of trust is "the comfort to make yourself vulnerable." Social scientists have formulated many more definitions, types, and levels of trust. For example, one level of trust is knowing what the other will do. Some call this "cognitive trust", a belief about the other's reliability. But an even higher level of trust is allowing the other side to act in your place. Complicating matters further, some individuals and cultures trust quickly (Americans are said to be in this category), other cultures more slowly.

Culture clash

Geert Hofstede, the famed culture scholar, calls culture "the programming of the mind that separates one group from another."[11] Culture defines each person's principles, values, beliefs, and behaviors, including communication behavior. As a result, in any cross-cultural communication, the receiver is more likely to misinterpret messages or cues. Hence, the familiar complaint of miscommunication across cultures. Worse, small cultural gaffes lead to culture clashes, mistrust, and eventually conflict. In fact, culture is so important in offshoring that we devote an entire chapter to the topic (Chapter 9).

Conclusions

Each of these five centrifugal forces lead to problems, and, in some cases, failed projects. The most difficult outcomes were noted by Jim Herbsleb, one of the pioneers in the study of global software teams.[12] He recently summarized several years of studying these kinds of teams. He concluded that there were two particularly serious outcomes. First, these dispersed teams cannot deal with unexpected events. Second, these teams suffer from issue resolution paralysis, which means that it is difficult for them to arrive at closure on difficult issues. Both outcomes stem from combinations of the centrifugal forces presented above.

Formalize and informalize

The five centrifugal forces (the five core problems of distance and time) can be mitigated by applying a wide range of organizational and technological solutions. In this section, and in the rest of this chapter, we present these solutions.

First, the dictum that should guide you is:

> *Formalize much of what is often informal*
> *and*
> *put more effort into creating informalisms.*

In other words, effective distributed collaboration requires attention to both of these actions simultaneously. *Formalisms* are the conventions, structures, and social agreements that standardize communication.[13] By formalizing, we mean that you inspect your informal work behaviors, and formalize them. Formalisms reduce expensive trial-and-error that is the basis of our natural coordination. The other side of the coin is to intentionally create *informalisms* (a truly informal word!). Creating an informalism across distance means deliberately creating social relationships between distant individuals. After all, distant collaborators who have some kind of social relationship perform better on a common project.

In this section, we present eight practical principles of formalisms and informalisms which appear in Table 8.1 and are described below.

Table 8.1 Formalisms and Informalisms for more effective distributed coordination

Formalisms	*Create a rhythm of interaction*, such as weekly meetings in real-time.
	Iterate for synchronization, with frequent deliverables across distance.
	Standardize communication protocols.
	Build an awareness infrastructure.
	Create protocols for acknowledgments and urgency.
Informalisms	*Create a cohesive team culture* by fostering relationships across distance.
	Foster interaction via real-time interaction.
	Put warmth into cold e-mail by the taking the time to create e-touch.

Eight practical principles

Formalize! Create a rhythm of interaction

Rather than rely only on one-to-one interactions between distant collaborators, managers need to create a rhythm of high interaction through meetings that synchronize, coordinate, and create a regular *rhythm* to the project.[14] This is an unusual recommendation since most management books advise to minimize meetings. Meetings have become so vilified as to become a humorous cliché. However, in distributed collaboration, meetings between individuals who are far away are a vital means for bridging the problems of distance. These meetings force *interaction* among the distant participants.

These meetings need to be in real time: they should be via telephone, or video, or web-conferencing. How frequent should these "dreaded" meetings be? The default, as seasoned managers know, is weekly. But daily meetings may be possible: one successful Russian–Swiss distributed software project conducted a strict, daily 15-minute tele-meeting at the beginning of the work day to kick-off the day's work. This project stayed synchronized. (This was possible because the sites were two time zones apart.)

Formalize! Iterate for synchronization

Distributed collaboration needs to formalize frequent synchronizations. By this we mean setting up work so that there are many small iterations, also called deliverables, that are sent off from one distant partner to another. Why is this important? Because nearly all the informal means that we naturally rely on are no longer available over distance: face-to-face meetings, or spontaneous conversations, such as a quick dialogue in the hall.

The answer is to iterate frequently. Frequent iteration (synchronization) addresses fundamental coordination problems of work over distance, and is likely to discover a problem before it is too late and delays the task at hand. Project managers need to introduce many small deliverables into the project timeline. The Work Breakdown Structure,[15] used by most project managers, should formalize these frequent iterations.

The right number of iterations is dependent on the context (task, people, and size). If iterations are too frequent (too granular), they overwhelm those who need to verify or add to them. But, iterations of once a week should be a minimum frequency to allow true synchronization between distant partners. In a similar spirit, these iterations are called "sprints" within the Agile development methodologies, which are described at the end of this section (and, according to the Agile methodologists, these "sprints" should certainly be no longer than two weeks).

The deliverable itself may be any work object (complete or incomplete) including: plans, outlines, prototypes, simulations, design reviews, test results, software code reviews, module integration, and documents. And, each of these deliverables should have a predefined recipient and predefined verifier.

Formalize! Standardize communication protocols

Most individuals in distributed collaboration groups are inundated with e-mail messages. Research has validated what we already know,[16] that we are distracted and less effective with the hundreds of telephone, e-mail, and Instant Messaging (IM) messages we receive daily.[17] Disorganized work groups tend to send each other too many messages with too much superfluous information. Now, more than a decade into the e-mail age, there are some who ignore e-mail because they have become overwhelmed. Some managers have gone so far as to ban e-mail.[18] Furthermore, in the software development cycle, reliance on e-mail loses critical design rationale that is buried in personal e-mail folders.

Since e-mail messages lead to information overload, they need to be reduced, standardized and replaced with workflow tools and project repositories. The overall goal should be to migrate most project-related information into repositories and databases so that the individual can *pull* them as needed.[19] Some of the information flowing in e-mail or IM also needs to be captured in these repositories, which is rarely the case today, since these messages are either stored in personal folders or not stored at all. Thus, persistent IM (in which all messages are stored) is a partial solution: one that is beginning to appear in the marketplace.[20]

Terminology is also part of the communication protocol, and needs to be standardized and formalized, since teams at different sites inevitably have their own interpretations for what things mean. Therefore it is best to agree on a common terminology upfront, particularly about methodological phases. Then, document these terms in the common repository.

Formalize! Build an awareness infrastructure

We operate more effectively as a group when distant collaborators are aware of all the dynamic factors that affect each other's work.[21] Recall that lack of awareness was introduced in the previous section as a one of the reasons for the "coordination breakdown" centrifugal force.

Research on awareness has its roots in the US military with the goal of giving a commanding officer a feel for what is happening in the battlefield, all this with the aid of technology. The US military researchers labeled this *situation awareness.* In subsequent years, numerous researchers have parsed out situation awareness into a number of important sub-types which are relevant to distributed collaboration (see Table 8.2).

More distributed software teams are creating "White Pages" that include: each person's name, contact information, and availability information ("I am available 09:00–18:00 GMT Monday–Friday"). It is useful to turn these "White Pages" into "Yellow Pages" with full information about each person's expertise, history, and tasks. This is a part of an *expertise search tool* that large companies have been adopting.

Formalize! Create protocols for acknowledgments and urgency

A protocol is a code prescribing adherence to correct etiquette. Why bother with etiquette? When collaboration crosses many time zones and relies on asynchronous e-mail, we do not know if our partner received our questions or requests. Instead of an

Table 8.2 Awareness types[22]

Awareness types	Answers the question	Example	Techniques to increase this type of awareness
Activity awareness Task awareness	Who's working on which task? Who has done something and where is it?	Has Hans finished the interface module, yet? I need to start testing its impact on my program.	• Repositories. • Project/task systems. • Status meetings. • Pairing across sites.
Process awareness	What piece fits where? How does Joe's task fit into my task? What to do next?	I just learned something important about the memory module. Now, who needs to know about this in the project?	• Organizational charts. • Expertise search tools. • Workflow tools. • Integrated development environments.
Availability awareness Presence awareness	Who's there to answer a question? Who is around?	June Hee is gone today, who can tell me if the cross-site meeting will take place?	• Instant Messaging. • Team calendars. • Individual availability schedules.
Environmental awareness	How does Natasha's work environment (weather, office, commute) affect her? Is it dark in London now?	The weather this morning was awful disrupting the morning commute. There is a general strike in the country, so daily life is disrupted. Joe has to go to another room to access the internet.	• Team/site dashboards with context, environmental data, news. • Etiquette that includes some exchange of environmental information. • Exchange photos of workplace. • Desktop video meetings.

acknowledgement, we often hear silence. The days tick by. Are they working on our request? Silence across distance leads to all kinds of interpretations, most of which are unhealthy and eventually lead to a spiral of mistrust.[23] Instead of silence, distributed collaborators need to establish a protocol of immediate acknowledgement. An acknowledgement does not necessarily imply task completion, but rather a message in the following spirit: "I received your message inquiring about the procurement interface and will reply with an explanation by Friday, July 12th."

Urgency also needs a protocol since it is relative: different people assign different priorities to messages and tasks. In some way, urgency (or lack of urgency) needs to be expressed across distance. Urgency can be addressed via escalation protocols intertwined in the managerial chain of command: e-mail to telephone, to cell phone, to home telephone (for very urgent items).

Informalize! Create a cohesive team culture

Recall that one of the five centrifugal forces is that distributed groups are rarely cohesive. However, effective managers can take some steps to create cohesion, to create a sense of a common team culture. This will help ease communication and foster trust across distance. We describe several techniques to do this here.

Developers tasked to work together over distance do not have the luxury of taking time to slowly build trust but must forge *swift trust* at the onset of their working relationship. Swift trust can be achieved by highlighting the reputation and professional qualifications of team members. Technical people tend to respect (meaning trust) other individuals who they view as technically qualified. Therefore, take the time to explain the pedigree of the distant partner: he was educated at the selective Indian Institute of Technology; she worked for the ground-breaking Irish firm, Iona.

A preferred, though expensive, means of building trust early in a project is to bring everyone together for a face-to-face kick-off meeting. We human beings form into cohesive groups when we develop common (informal) experiences together. The common experiences can be eating together, drinking together, playing together, or just chatting with one another face-to-face. Jim Herbsleb[24] speculates that one of the missing ingredients to effective distributed teamwork is the informalism of "goofing around." Of course, collective kick-off events are more common during times of company prosperity and for high visibility projects. Lower-cost options include well-orchestrated, virtual kick-off events enabled with high-resolution video-conferencing.

Other common experiences require creativity and enthusiasm. Some distributed software teams have created online team discussions, and given them a fancy label, such as "virtual café," or "virtual retreat." These common activities can be asynchronous or, perhaps, once a month in real-time. Activities can vary from solving a public policy problem, to discussing the football World Cup, or even watching a football game at the same time. Researchers have validated these recipes: virtual teams that have more social communication achieve higher levels of trust.[25]

Informalize! Encourage interaction via real-time communication

I make sure to do a synchronous conversation (IM, phone, video, and face-to-face) with each person on my team at least once a week. I even have a list of all team members and check it off. ☑

A European Project Manager

These days, instead of Management By Walking Around (MBWA), we do MBFA (Management By Flying Around).

An American Project Manager

Since human beings are more effective with real-time interaction, then any type of real-time communication is preferred to relying strictly on e-mail. Like the European Project Manager quoted just above, consciously shift to more real-time channels. Even paired-programming (advocated by "Extreme Programmers") can be conducted with a headset telephone so one programmer can be keyboarding while speaking, as the programmers share their computer screen using a collaboration tool, such as

Netmeeting (we come back to some lessons from Extreme Programming at the end of this section).

Off course, the most effective real-time communication is *face-to-face* and therefore there should be regular travel between the project sites. The *liaison* is the label we use for the individuals interacting frequently with distant sites. They are the channels through which synchronization is conducted and through which messages are transmitted.[26] These liaisons use not only face-to-face communication, but augment this with other real-time channels, such as telephone, or video. The most effective liaison is often the expatriate linking the organization to his birthplace. In that role he is not only a liaison, but a true ambassador. Liaisons are nurtured via rotations (of weeks or months), or so-called "onshore presence" (rotating lower-wage professionals to be near their clients or partners).

In spite of the benefits of face-to-face interaction across distance, the travel budget is one of the first items to be cut from distributed projects when budgets become tight. Sometimes this cost-cutting ends up costing more. One study found that cross-site trust is low unless at least *one* person made at least *one* trip between the software development sites.[27]

Informalize! Put warmth into cold e-mail

"I greet you and I invite all of you to my home for the dinner this evening.
It's just after the corner of one of the main street of Rotterdam. ;)))"

A student e-mail from a global virtual team exercise[28]

Note some interesting attributes of this message. This message puts the writer's colleagues in context by giving them a sense of where he lives. The message has at least three grammatical mistakes, but the reader probably does not care because the message is warm, welcoming, and its fantasy of a cozy dinner in his home adds to the warmth.

We already noted how easily messages can be misconstrued and lead to bitterness. A bit of warmth can do wonders to alleviate distance. In general, this is about building *e-touch* over the net. Some early research shows that those that are good at building social capital in the face-to-face world are also good at doing so over the net. Here are some useful e-mail clauses to incorporate into your next messages:[29]

- "… thank you for your flexibility in working with us on these points …"
- "… we have been making great progress on this …"

Lessons from "unconventional" distributed approaches

Two of the most radical and successful software movements have also fused the informal with the formal to overcome problems of distributed collaboration. The two are the *open source software* (OSS) community, best known for Linux, and, separately, the *Agile Software Development* movement, best known for Extreme Programming.

How has the OSS community managed to develop robust software in completely distributed collaboration? After all, OSS projects are born distributed and rarely perform any co-located or synchronous activities.

Walt Scacchi found several types of informalisms that make this type of distributed collaboration succeed.[30] Like other informal processes, requirements are co-mingled with later phases, with design, coding, testing, and documentation. The requirements are organized around persistent, globally accessible tools and repositories: websites, site content directories, source code directories, threaded e-mail, bulletin board forums, bug and enhancement descriptions, and version descriptions. All this allows those in the distributed community to trace the development and evolution of the project and its design. These are examples of the formalism principles that we noted earlier: standardizing communication protocols and building an awareness infrastructure.

The OSS community also standardizes around a rich set of Internet-based tools: SourceForge, Bugzilla, forums, listservs, newsgroups, and IM. In lieu of formal requirements or any face-to-face interaction, the community develops shared understanding using screen-shots, guided tools, "how to" guides, and execution scripts.

Extreme Programming (abbreviated as XP) is the best known of the Agile methodologies. Agile methodologies emerged as a reaction to the "heavy" methodologies that are typified by CMM. XP believers (and they are believers!) advocate working in small co-located teams. Programmers are paired with each other, working side by side, helping, guiding, and mentoring each other. This is a "high contact" approach with much face-to-face interaction. Therefore, XP is the *antithesis* of software collaboration over distance.

Nevertheless, XP's advocates have learned to adapt it to the reality of work over distance: there is even a Dispersed XP (DXP). They learned a number of lessons,[31] many of which were discussed in this chapter: start the team in a single location (to get to know one another); designate travelers that physically move around sites (this is similar to our label of "liaisons"); agree to a block of common time (this is similar to the first formalism principle of a rhythm of real-time interaction); agree on common communication tools; and agree to a common set of coding guidelines, coding styles, and modeling guidelines (we called these: standardizing communication protocols).

Managing time differences

"I can have a high priority, but at the time when [the US colleagues] are sleeping they won't answer me. [...] Same thing with me: When I'm not working, I'm not reading my mail."
                              Singaporean computer engineer working on a distributed team with
                                                              American partners[32]

One of the biggest factors in the amount of time it took us to fix bugs was the time zone difference. When we diagnosed a bug, we tried to determine who in India, Europe, or the US needed to look at it. We'd reassign it, but if it had to go to someone in Europe, or India, there would be a delay of several hours before the assigned person would be in on their regular work shift to look at it. If it got assigned to the wrong person, they might reassign it in turn with another potential time zone delay. With problems that were difficult to diagnose, the ownership of the bug could pass back and forth, with time zone delays adding up to several days of little or no work getting done.

Software engineer, Siemens, USA

Time differences are a constant nuisance when offshoring. Time differences are more than merely time zones, they also include different starting and ending times at work, different religious and national holidays, different weekends, and different lunch and other break hours. Here we present these time differences and some of their related best practices, based on research Carmel conducted with Alberto Espinosa.[33]

Managing time-zone differences

Experienced global managers have a bag of tricks that they use implicitly, almost intuitively, in managing and coordinating across time differences. It is quite clear that distributed teams do not treat time differences as static, but rather adjust and adapt to them. The 10.5-hour difference between New York and Chennai (India) is not "fixed" in that sense. These tactics are summarized into three categories in Table 8.3 and we explain each of them below.

Asynchronous tactics

Effective distributed teams learn to conduct as much work as possible in non-overlapping time. This translates into a number of tactics. First, effective teams *formalize* and *structure* activities and messages so that they convey information in a more effective manner, thus

Table 8.3 Tactics to overcome time differences[34]

Category	Tactic
Asynchronous (non-overlap)	• Structure and formalize into workflow tools. • Plan the work day: bunch-and-batch; plan dialogue for overlap window.
Synchronous (overlap)	• Enlarge overlap window by working longer. • Enlarge overlap window by shifting work hours. • Enlarge overlap window by always being available. • Enlarge overlap window by creating a 2nd shift in the low-cost offshore destination. • Create individual liaison roles who adjust/enlarge their own hours rather than the entire team's. • Create fixed daily, or once-a-week, overlap periods between sites. • Synchronize individuals who are working closely together (in paired tasks). • Break the e-mail chain.
Awareness	• Reminders and coaching. • Easy access to current time, calendar, and holiday schedule of distant individuals.

reducing the interaction required by a need for clarification. Distributed teams become "bureaucratic" in other ways: they carefully define the collaboration workflow, tasks, owners, and deliverables in order to reduce the need to coordinate via real-time interaction.

Time-separated collaborators learn through experience to plan and organize their work days to maximize any overlap window for real-time dialogue, such as teleconference meetings to resolve problems. They also learn to bunch-and-batch their work to be delivered to the distant sites at the end of their day.

Many of these tactics also hint at a hidden benefit of time differences: interruptions are reduced. There are less telephone calls, meetings, and instant message requests. People can concentrate on "getting their work done."

Synchronous tactics

The richer set of tactics are those that use overlap time (and hence *synchronous*). Most familiar, teams tend to enlarge the overlap window by shifting and expanding work hours. For example, European staff may start late and work late, so as to have greater overlap with their American counterparts. Conversely, the Americans may start early, either everyday, or at least on some weekdays, so as to expand the overlap time with their European counterparts. Many of the new offshore companies are staffed with young, ambitious software engineers that work long hours, often late into the night, creating overlap windows with Europe and America.

In practice, when collaboration crosses many time zones, time window expansion is practiced by only some of the distributed team, particularly the managers, team leaders, and liaisons. Liaisons help team members interact across sites. For example, in one case a large software team with members in Britain, Germany and India trained Indian software engineers in Europe for several months to serve as liaison engineers.[35] These liaison engineers returned to India and then adjusted their work schedules to increase their overlap window with their British and German counterparts.

One time zone "trick" is to know how to "break the e-mail chain." The e-mail chain begins when, in time-separated asynchronous communication, one engineer initiates a message; the receiver, on the other side of the globe, does not understand it fully and asks for clarification; the original sender attempts to clarify; the receiver then interprets it incorrectly and responds accordingly; the receiver then sends another clarification. Meanwhile, an entire week has gone by. Experienced engineers stop this chain early "by picking up the phone" to clarify the message.

Awareness tactics

This last of the tactical categories is targeted at younger, less experienced team members. They are not used to thinking about their counterparts being asleep while they are working. They are not used to computing the direction of the time difference ("Is it $+7$ hours or -7 hours?"). They cannot recall when their counterparts shift to daylight savings time (it is at different times in different nations). Small reminders and coaching help address this problem. The distant individual reminds her counterpart that the scheduled meeting is set for 06:00 local time. A simple tactic is to post hours and time differences on the common team website.

The soft costs of time-zone differences

Many of the tactics described above are based on time-shifting. Time-shifting takes a toll on individuals by rupturing the boundaries between work and home life. We all hear from software engineers spending many evenings, nights, and early mornings, in telephone conversations across the oceans. With the ubiquity of mobile telephones and other wireless gadgets, key individuals are always reachable – any day, any time, anywhere. Balanced teams try to rotate meeting times in order to shift the burden of late-night (or early-morning) conference calls. But, all too frequently the dominant site (headquarters) dictates meeting times convenient to their normal work day. One familiar result is the team member who regularly falls asleep in the middle of teleconferencing meetings.

Since many of these tactics involve some personal inconvenience, not everyone plays along. One collaboration between a British and California office (8 hours apart) was at a major American technology firm. During one project phase the team was working on urgent software fixes. The British technical expert insisted on starting the day early, while his California counterparts liked coming into the office late and working late. Neither side made accommodations, resulting in no overlap window for synchronous communication. All communication with the British expert relied on one e-mail batch per day which "really slowed down the work."

Other time differences

While less dramatic, other time differences impede collaboration. We describe them here.

Work hours

Daily work hours vary by country. Here is a sample of such norms. In India, formal work hours for technology firms are from 09:30–18:00, but "no one goes home at 18:00." In China, traditional work hours are from 08:00–18:00 with a two-hour lunch break. Huawei, China's most successful technology company, has kept the two-hour lunch tradition, but most other technology firms have moved to 09:00–18:00 with a one-hour lunch. In the USA, the work hours are 09:00–17:30 with most lunches beginning at 12:00; in Spain, workers begin late in the day, have longer and later lunch breaks, and finish their work day often much later than 19:00. Americans are more likely to plunge into work as soon as they get to the office; the French (who also tend to begin the work day late) tend to spend the first period of their office time drinking coffee and in small talk with co-workers.

Breaks

Breaks, such as lunch breaks, disrupt overlap windows. A European distributed software team at Lucent found that even a one-hour time-zone difference between two sites substantially affected the team's ability to communicate interactively because it reduced their overlapping time by 4 hours: one hour at the beginning of the day, one hour at the end of the day, and one hour during each site's lunch break.[36]

Weekends

While much of the world has standardized on a two-day weekend on Saturday and Sunday, this is not universal. In Hong Kong, the work week is Monday through Friday plus a half day on Saturday. In Arab countries weekends revolve around Friday (their Sabbath) with an additional day off either on Thursday or Saturday. In Israel, the weekend is Friday and Saturday. For Americans working with Israeli partners, the Israeli weekend creates a long "blackout period." The Israelis have essentially left for the weekend when the Americans come to work on Thursday morning. The reverse happens with the Israelis, who work for much of the first two days of the week without being able to contact their American colleagues.

Holidays

The patchwork of national holidays is bewildering. One American technology firm had staff in more than a dozen European nations. The Human Resource (HR) director reviewed the calendar for all these nations and came to the remarkable conclusion that due to different, non-overlapping national holidays, there are only 50 regular works days in common in any given year for the purpose of scheduling synchronized meetings (e.g. the entire month of August is not usable in many European nations).

Collaborative technology

No reader of this chapter is unfamiliar with the menu of collaborative technologies available for distributed collaboration that appears in Table 8.4. The right mix of these technologies enables effective distributed software collaboration. Too often, though, distributed collaboration tends to rely too much on e-mail, and too little on other technologies.

Table 8.4 Types of collaborative technology

Asynchronous technologies	Synchronous technologies
E-mail	Voice telephony/Internet telephony
Voice mail/video mail	Audio-conferencing
Online discussion groups	Video-conferencing (meeting room)
Calendaring	Video-conferencing (desktop/web-based)
Collaborative authoring and commenting	Web-audio hybrid meetings
Project Management	IM (Instant Messaging)
Production tools and repositories such as configuration management systems, issue tracking systems, workflow tools, knowledge management	Whiteboard/Screen Sharing

Some people claim that the collaborative technology panacea is video-conferencing. So, let us examine this assertion. It rests on the premise that "rich" multi-sensory, interactive communication is more effective. This is called the theory of *media richness*.[37] The implication is that we tend to reach out and use "richer" media if we can. Accordingly, the richest technologies are synchronous (real-time) technologies; and the richest of these is video-conferencing. Video-conferencing is the closest we have, these days, to face-to-face communication. Indeed, more and more software professionals are using video-conferencing, at least on occasion, for meetings, or better yet, for more informal discussions over distance.

As video-conferencing continues to improve, the goal is *telemersion*. In this form of virtual reality, you will work surrounded by several cameras capturing your every movement, while viewing 3D images and listening to audio in 3D through surround-sound. At one end of your office area, near the espresso machine, will be a *live-wall* (or video-wall) in which you will be able to walk up and spontaneously chat with your Indian colleague, an ocean-away, who just walked over to her office's juice bar. She introduces herself: "Nice to meet you. My name is Sudha, and I joined the team here as a quality assurance specialist this week. I look forward to working with you."

But fantasies about video-conferencing tend to clash with some realities. Many software professionals do not like working with real-time video for a host of reasons: they are not sure about the correct etiquette of dealing with someone far away, they are afraid that their interaction is being monitored, they are shy, they really do not see why it is helpful when so much can be done using e-mail. And, of course, the interoperability and usability issues of video-conferencing have not been solved.[38]

There is now better appreciation for the advantages and limitations of all of these collaborative technologies of Table 8.4. Social scientists have been studying their impact on work groups and developing a better understanding of the interplay of new features with behavioral reactions.[39] Tom Erickson, of IBM Watson Research Center, led a group that created a persistent chat tool for distributed teams with the intent of fostering greater community among software developers ("persistent" means that the text is stored and does not scroll away). The tool was introduced to two distributed teams, one in the US, and the other distributed globally. Each team used the tool differently. The American project used the tool as it was intended, that is to generate a sense of community. On the other hand, the global project only used the tool in the hectic period just before release. Erickson noted two adaptation differences. First, the time zone differences do not allow much real-time communication. Second, the global project had almost no person-to-person history across sites, so it was socially more difficult to use real-time communication.

Companies can benefit from collaborative technologies by investing in a *mix*: instant messaging, video-conferencing, *high-quality* audio-conferencing services, *rich* application-sharing environments, group calendars, knowledge management systems,[40] and integrated repositories.

On the softer side, this technology mix needs to be layered with support people: dedicated collaborative technology specialists who tinker and customize the tools, train the users, and help teams to make the tools work. The Tier-1 IT providers have all invested in creating their own collaboration suites; sewing together off-the-shelf technologies with some home-grown features.

We have yet to reach the holy grail of collaborative technology: the all-in-one, seamlessly integrated suite of tools that are fully interoperable (and secure). The collaborative technologies of the future will likely be composed of three integrated parts:[41] shared workspaces, high-quality video-conferencing, and comprehensive event capture repositories. The Intel case study, below, describes a collaboration vision for the company's future.

Case study Intel's vision for new collaboration technologies

Intel is one of the most globally dispersed high-tech firms, with research and development (R&D) facilities in 10 countries (Ireland, Israel, USA, Russia, India, Malaysia, the Philippines, China, UK, and Spain), 11 fabrication plants, and 4 assembly plants worldwide. In total, Intel has 236 offices around the world.

The collaboration landscape is complex: manufacturing, design and software engineering, as well as nearly every corporate activity, are all conducted with distributed teams. Telecommuting is encouraged and quite common at Intel. At the same time, some of Intel's facilities, in a handful of countries, do not even have high-bandwidth communications.

In spite, or perhaps because of its prominence, wealth, success, and history of dispersion, Intel still found problems with its collaboration environment. Indeed, the very fact that the company depends upon remote teaming for daily productivity raises the bar for collaboration tools.

An internal study[42] conducted in 2003 showed just how "virtual" Intel had become:

- On a weekly basis Intel was conducting 8300 web-based collaborations per week and 19,000 audio bridges.[43]
- 51% of employees regularly worked with others who used different work processes.
- 40% worked with people who used different collaboration tools.
- 71% of employees collaborated with people who speak other languages.
- A long-ignored problem was uncovered and measured for the first time: *Multi-teaming*. Almost two-thirds of Intel employees were members of more than three teams!

Most intriguing was the finding that distance, in and of itself, did not impact (perceived) team performance, but that the confusing myriad of Intel's collaboration tools negatively impacted team performance. Employees were complaining about too many tools that do not inter-operate well.

The Virtual Collaboration Research Team (VCRT) was formed in 2002 to address these problems by creating a collaboration *vision* for Intel. At large companies such committees often create reports which can easily end up ignored on a shelf. The VCRT set out to influence management thinking by continuously communicating the vision upwards, and gaining support to link the vision to mainstream plans, turning those ideas into practice step by step.

The VCRT was composed of individuals from various internal groups. These were people who were passionate about collaboration. This is a "rambunctious team" as one member described it, "full of energetic and creative thinkers." VCRT had a nucleus of experts who were involved in influencing collaboration in their "real jobs." For example, one of the members was the Collaboration Architect within Intel's corporate IT Group.

The vision

After extensive deliberations, and drawing on some prior work at Intel, VCRT's ambitious vision coalesced. Specifically, VCRT decided that collaboration must focus on two thrusts: multi-teaming and work across time differences. First, support for multi-teaming means that all of the employee's projects and tasks must be on the same desktop. Cross-project activities need to be arranged into composite views. Second, in order to address the "time warp" problem, of multiple time zones, Intel employees would be able to join "the meeting" asynchronously, whenever they wished. The vision also included a "desired user experience" as its organizing principle: interoperability of applications for seamless, one-click navigation and transactions. Finally, given that most global team members at Intel never meet face-to-face, the vision includes some kind of socializing interface to allow *expressivity* in a multi-cultural environment.

The components of this vision were then formulated into a layered architecture. At the top layer is the individual workspace of each Intel employee. In the layer below that is the core architecture made up of four tool clusters:

1 *Co-authoring*, including document and web content, software coding, and inline annotations.
2 *Project Management*, including task tracking, issue tracking, and resource management.
3 *Team Management*, including team building, discussions, and team member data.
4 *Meeting Management*, including calendaring, meeting structures and workflow, and conferencing.

The last cluster of functionality became the most radical in its social engineering goals and the most controversial within the VCRT: to get Intel employees to change behaviors by expanding the notion of the meeting. In essence VCRT were redefining the definition of a meeting. In this vision employees would do more tasks before the meeting and after the meeting, or even instead of a real-time meeting. Each Intel team will drive more and more structured activities into pre- and post-real-time meeting sessions through an asynchronous meeting participation mode that spans each team member's range of available working hours. Activities will be completed anywhere within the time window that the (redefined) "meeting" takes place.

Individual team members will be assigned activities with pre-conditions and deadlines. An intelligent workflow scheduler will suggest possible open timeslots that will allow individuals to drag and drop the activities onto convenient timeslots in their calendar. Many collaboration activities lend themselves to such asynchronous meetings: preparing assigned sections of content for documents; document review and annotation; as well as gathering and summarizing information that the team needs to take into consideration.

Team members will see everyone's progress towards completion of the asynchronous activities through a status monitoring view. Personalized reminders will go out automatically relative to each individual's time zone and calendar prior to each employee's own settings for the task start time.

While VCRT's focused on transforming the notion of the "meeting" into asynchronous activities, it did not preclude real time interaction. For example, if two team members will happen to notice that they are both online in the asynchronous team workspace they will be able to easily switch to an *ad hoc*, real-time meeting for a quick real time dialog. The more traditional real-time meetings will be scheduled for a specific timeslot and will be supported by structured tools. But even though these real-time meetings will be structured, participants will be able to switch on any *ad hoc* elements they wish during the session, such as deciding to invoke data sharing with a single click that joins everyone together immediately.

Influencing the many stakeholders

VCRT members understood the difficulty of communicating a new collaboration *vision* into mere words on a page. After all, everything sounded to others like some existing collaboration environment. Rather than create text to capture the vision, they created an illustrative four-minute demo video, using multimedia and a professional announcer.

The demo video showed the Intel employee of the future utilizing a 3D integrated, multi-perspective desk-top and collaboration environment. The audience sees the employee's screen with workspace, colleagues, timeline, work products, and meeting lists. A project that requires immediate response is blinking. To address the need to support the Intel reality of multi-teaming, the demo shows how the employee can see all his responsibilities for all his teams on one timeline. He can drag and drop any of them into convenient timeslots.

The inspiring animated demo and the compelling virtual survey data became tools for changing the corporate vision. It was a vision that resonated with Intel

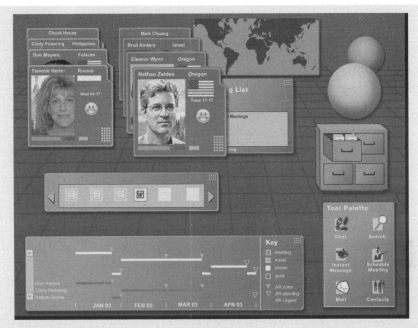

executives, particularly those whose divisions suffered most from time zone differences. These executives became most excited about the collaboration vision. The VCRT challenge was to create momentum for change from managers who had many other objectives. In fact, it was only after VCRT's vision gained acceptance outside of their home base in the IT Group that their cause was taken seriously.

From the beginning VCRT saw their mission to influence four audiences rather than just one. Of course, the first was to influence collaborative work inside Intel including all business units and functions. The second was to influence the software marketplace to support their vision. In particular, leading software product firms in office products and collaboration tools were key targets. The impressive demo became a visual means to convey the vision to these critical firms. Multi-teaming was not supported adequately by any product on the market. Instead, all current tools assumed a fairly traditional organizational view of one hierarchy, one project per person, with no matrixing. VCRT emphasized that this needs attention.

Third, VCRT wanted to influence the marketplace for Intel's own high-speed chips as well as chip design. By understanding the needs of high-end collaboration Intel chip architects could facilitate the needs of the future generations of collaboration platforms. These platforms will include support for rich media and parallel processing in order to be responsive to the social and multi-tasking requirements of collaboration tools. Finally, VCRT wanted to influence Intel's strategy in investments and alliances by directing the company towards firms with promising collaboration solutions.

By 2004, VCRT could boast of a few successes. The latest Intel long-range planning document included the VCRT vision as part of the planning process. VCRT's vision was now embedded in management's conversation and, as is the barometer these days, into management's slide shows. It was also visible within an internal Intel program called eWorkforce which supports knowledge work throughout the organization. VCRT applied for a US patent on their collaboration concept. The Virtuality Study became an annual tracking tool that may also be turned into a corporate benchmarking tool. Outside industry analysts featured VCRT's vision in industry gatherings. These analysts pointed to Intel as a global corporation that rigorously identified its needs as a globally distributed company and as one that, therefore, will be a trendsetter in collaboration solutions.

Selecting the right people for distributed collaboration

Those who know how to create social capital face-to-face are often those who know how to create social capital online.[44]

Beyond technical and managerial skills, selecting individuals for distributed software development should also be based on their ability to work over distance.

There is now a set of best practices in choosing desirable skills and talents for distributed collaboration.[45] First, the individual needs to possess good communication skills, such that she can not only send a message which is understood but can solicit feedback across distance. Some empathy is needed to head off the quick spiral of anger-mail and bitterness that may emerge via e-mail exchanges. A history of using a mix of multiple technologies (e.g. teleconferencing, video-conferencing, workflow systems, team repositories) is essential. Since there is more isolation in distributed work, individuals need to be capable of self-management.[46] Because of cultural differences, heightened cultural awareness is needed. For example, an Asian team member remained silent when he disagreed. Will your Western manager be aware that this was done in order to save face? Will he know to interpret this silence and to politely prod further for points of disagreement?

Leaders are especially handicapped when they become *virtual* leaders because their success was achieved through face-to-face contact. In order to compensate, many virtual leaders practice Management By Flying Around (MBFA). Thus, one necessary condition for a virtual leader is tolerance for frequent flying.[47] MBFA is essential for most offshore projects.

Conversely, too many virtual managers resort to management by e-mailing around (MBEA), or managing by (ever more detailed) contract, or managing by milestone. These approaches may work for straightforward projects with experienced staff, but fail in other instances. Asking a remote team to get a task done by a certain deadline is not nearly as effective as flying there to make sure that they get it done on time.

Undoubtedly, the virtual leader cannot rely solely on MBFA because of time and expense. Using a mix of technology channels he must communicate his messages: inspire his engineers; convey and then reiterate the project vision; become the team "glue" by reminding people who is doing what, and who needs something, and when. Successful leaders also coach their subordinates. In sum, virtual leadership is an even more demanding position than leadership in the traditional co-located work environment.

From the HR management perspective, the ability to work over distance needs to enter into the long list of qualities that the company seeks in hiring and promoting technical staff. HR staff can also enhance these skills with training courses in group processes and in cross-cultural communication.

Distance considerations in organizational design

Some companies resist distance *by design*

> *Microsoft, the icon of the software industry, has long resisted distributed software work, preferring to concentrate its R&D work in its large Redmond campus in the Northwestern US. Microsoft is also keenly aware of distance within its campus, frequently moving software engineers within floors and buildings to create proximity.*

> *Perhaps sometimes Microsoft goes too far. In 2003, it acquired a California company called Placeware that creates software to help distributed workgroups collaborate over distance. At the time of the acquisition some inside Microsoft argued that precisely because of the nature of its products Placeware employees should be left in California and be "forced" to work over distance with headquarters 1000 km north in Redmond. But, Microsoft executives resisted and, as has long been the custom at Microsoft, Placeware offices in California were closed down and key employees were asked to relocate.*

Not all companies can or want to resist distance by design. Offshoring forces you to work over distance, with all its inherent difficulties. In offshore outsourcing, nodes of people involved in the software project are dispersed around the world: the client company is often dispersed in several locations, the outsourcing provider often has staff in several locations onshore and offshore. Most software product companies have distributed software R&D centers. Many projects are not only *distributed* across sites, but may even be *dispersed.* This means that far-flung individuals are assigned to the project who are working "alone" in some other office or may even be teleworking, isolated from the rest of the project clusters.

Overcoming distance and time implies that more attention needs to be devoted to organizational design. There are two principles of organizational design for distributed software work:

- *Principal I*: *Reduce the number of project locations as much as possible.* The number of project sites is also correlated with overall project size. Herbsleb and colleagues[48] found that the overall project team size matters; the single most important predictor of problems in distributed collaboration is the overall size of the project team.
- *Principal II*: *Reduce the dependencies as much as possible.* A dependency occurs when one location cannot make progress until another location finishes its work or otherwise solves a problem (whether that problem is large or small). By definition distributed collaboration demands that there be some dependencies.[49] In some cases collaboration will have highly *dependent* tasks. A better design is when tasks are quite *independent* between distant sites. In such cases, the interfaces between the sites are well defined (in other words the interfaces are "well architected"), so that the dependencies are minimized.

Clearly, high dependencies should be avoided, *by design*, because of the higher probability of breakdown, delay, and crisis due, in part, to miscommunications, language, and culture. Yet in some software collaborations one finds this type of organizational structure. Such structures may result from big events, such as mergers or joint ventures. The link between organizational structure and the software product architecture is summed up in "Conway's Law," which elegantly states that, in practice, the structure of the software system follows the structure of the organization that designs it.[50] Ideally, the opposite should hold.

In order to reduce the dependencies in distributed collaboration, there are a number of organizational designs for distributing tasks across distant sites (illustrated in Figure 8.3):[51]

- *By expertise*: Keep work that requires similar functional expertise in one place.
- *By product*: Each site works on its own components. The organizational structure follows the component architectures. Alternatively, the components are allocated by site.
- *By phase*: Keep entire processes in one place–design, code, test. Each of these phases ends with a hand-off to another site.
- *Satellite customization*: One site owns the core code for the product, while the secondary sites are involved in customizing features for each client near the client location.

Yet, even in these organizational designs, which reduce dependencies, the point of failure is often the dependency: the hand-off, or integration phase. These points of failure need the most proactive managerial attention.

We also note two other interesting designs to address some distance problems: *mirror* organizations and *onion layer* teams. A mirror organization is a symmetric organization at each of the distributed sites with identical structures and roles. This design makes it easy for a member of one site to identify his counterpart in another site. The paired

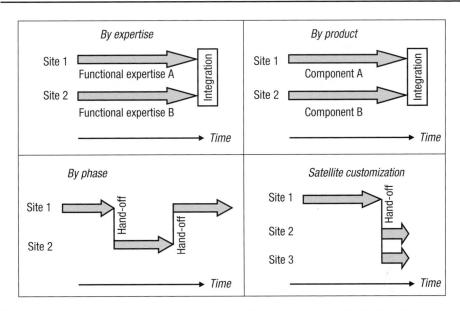

Figure 8.3 Four types of organizational design to minimize dependencies in distributed collaboration.

individuals communicate, develop a closer relationship and problem-solve together. The onion layer team appears in some Open Source teams.[52] In Open Source there are few developers that actually reside next to each other. Instead, a core team of three to four developers interact intensely with each other. These developers represent the core of the onion. Around them, in layers, are additional developers who review or modify code and contribute bug fixes.

In conclusion, while mirror and onion models are rare and have drawbacks, they point to the range of possibilities in organizational design to overcome distance. The important lesson is to be proactive about such design and not settle for inherited organizational structures.

Concluding lessons

- Overcoming the problems of distance require a mix of formalisms and informalisms: formalize things that have traditionally been done informally and put more effort into creating informalisms to nurture social relationships.
  - Create a rhythm of interaction between distant sites with regular real-time meetings.
  - Iterate for synchronization with frequent deliverables.
  - Standardize communication by shifting e-mails into workflow tools and repositories.
  - Build an awareness infrastructure.
  - Create protocols for acknowledgments and urgency.
  - Create a cohesive team culture by nurturing social relationships.
  - Foster interaction by encouraging real-time interaction.
  - Put warmth into cold e-mail by the taking the time to create e-touch.
- Overcome the problems of time through the following tactics:
  - Plan the work day using bunch-and-batch.
  - Enlarge the overlap window by working longer, shifting work hours, or always being available.
  - Create individual liaison roles who adjust/enlarge only their own hours.
  - Create regular overlap windows between sites.
  - Synchronize individuals who are working closely together (in paired tasks).
  - Break the e-mail chain by picking up the telephone.
- Incorporate distance into staffing decisions. Hire based on proven ability to communicate over distance and willingness to travel.
- All distributed collaboration teams need to invest in a rich mix of collaborative technologies: instant messaging, video-conferencing, high-quality audio-conferencing services, web-based conferencing, rich application-sharing environments, group calendars, workflow tools, knowledge management systems, and integrated repositories. The investments need to go beyond the assets and need to include specialists that customize the tools and train the users.
- Design the distributed organization to minimize dependencies across sites as much as possible. Manage the points of weakness — the hand-off points and the integration points.

9   Dealing with cross-cultural issues

With more and more software professionals working in distributed teams across cultures, four computer science professors decided to study the impact of cultural orientations on performance.[1] Computer science students from Texas and from Turkey were assigned into collaborative groups in a controlled experiment. Each of these virtual groups performed collaborative software design and programming. The professors evaluated the quality of the tasks at the end of the semester.

The researchers found that certain mixes of cultural orientations effected performance. If at least one of the members had a high power orientation the group's performance was less likely to be successful; if members had different destiny orientation scores the group's performance was less likely to be successful; finally, if at least one of the members had a high future orientation score the group's performance was more likely to be successful. (Each of these three orientations is explained in this chapter.)

Many software developers that we meet are relatively new to the topic of culture. They may have traveled and learned some of the superficial differences between some countries, such as greetings, but in order to become effective participants of global software development organizations, a deeper understanding is vital. The first two sections of this chapter serve as a mini-primer on this topic.

What is culture?

Culture, in the anthropological sense, according to Geert Hofstede, is the "collective programming of the mind distinguishing the members of one group... from another."[2]

Every adult is a member of many cultures. He is a member of an ethnic/national culture; she is a member of a religious culture; he is a member of a professional culture (such as a musician or architect or software engineer); she is a member of an organizational culture (such as Microsoft or Sony); and he is a member of one or more work groups and work teams, each with its own culture. Many of these cultural types, such as organizational culture and team culture, can be re-programmed in our brains fairly quickly (especially for those under 30 years old). However, *national culture* can not (and by this we include ethnic differences). The focus in this chapter, consistent with

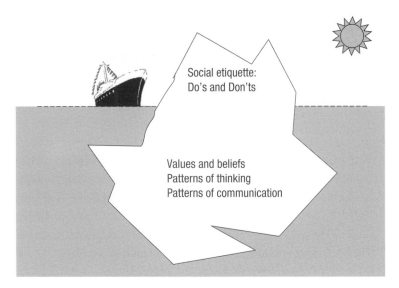

Social etiquette:
Do's and Don'ts

Values and beliefs
Patterns of thinking
Patterns of communication

Figure 9.1 Culture is like an iceberg, where most of it is "hidden beneath the water."

the rest of this book, is on distance and cross-border issues, and therefore this chapter emphasizes "cross-cultural communications."

Our national cultures are very deeply embedded in each of us, passed from generation to generation, and are largely programmed into us by the age of 10 years. Our cultural orientations manifest themselves in some behaviors and in phenomena that can be seen with our eyes. We may be able, with training and experience, to see some of these behaviors, such as body language, different decision-making norms, gestures, and business etiquette. Outsiders may learn some of the rituals, such as handshakes, or the correct protocol for answering a telephone call.

However, much of what is programmed into our individual culture is invisible, driven by deep values and beliefs which are very difficult to change or observe, as Figure 9.1 illustrates using the iceberg metaphor of culture (90% of an iceberg is submerged and cannot be seen). Values and beliefs include: good versus bad, ugly versus nice, dirty versus clean, and rational versus irrational. In every culture these dichotomies are interpreted differently.

Cultural orientations

While cultural radicals in academe now tend to scorn them, the seminal works of Geert Hofstede[3] and Edward Hall, who first defined key cultural orientations, have been essential to a generation of global travelers, business people, and virtual team members. In this section we summarize nine orientations formulated by Hofstede, Hall and other social scientists (there are still more cultural orientations and these can be found in many books about cross-cultural communication).[4]

Table 9.1 Power orientation index

Country	Power orientation index	Remarks
Israel	13	Hierarchy is less important
Germany	35	
The Netherlands	38	
USA	40	
Japan	54	
France	68	
Hong Kong	68	
India	77	
West Africa	77	
Indonesia	78	
China	80	
Russia	95	Hierarchy is very important

Source: Hofstede.

Power orientation

This is one of the most important orientations in the business context. It expresses one's emotional distance from subordinates and superiors. High-power orientation cultures tend to have more autocratic managers, while low power orientation cultures use participatory and consultative management styles. A subset of Hofstede's rankings for this orientation appears in Table 9.1. Individuals from high power orientation cultures are less likely to express disagreement with their managers, are less likely to be forthcoming online, and are more comfortable with an autocratic/paternalistic decision-making style. Managing in these cultures requires more authoritative communication. Since feedback in such cultures is not forthcoming, one has to develop informal relationships for feedback or, as we learned from numerous offshore collaborations with India, to train Indian employees to be more forthcoming (see the two case studies about India later in this chapter).

Relationship orientation

This is often referred to as *individualism* versus *collectivism*. This cultural orientation answers the question: How do you see yourself first and foremost – as an individual, or as part of a larger group? (see Table 9.2 for rankings). People from individualist cultures have a high desire for personal freedom, privacy, personal time, and personal challenges. They are expected to look out for themselves. There is higher regard for assertiveness and confrontation in work situations. There seems to be a strong correlation between wealth and individualism. Wealthy nations are much more individualistic, and as nations have become well-off their middle and upper classes rapidly assimilate individualistic orientations. Such is the case, with some qualifications, with India's new class of software professionals.

Table 9.2 Relationship orientation index

Country	Relationship orientation index	Remarks
USA	91	Highly individualistic
The Netherlands	80	
France	71	
Germany	67	
Israel	54	
Russia	50	
India	48	
Japan	46	
Hong Kong	25	
China	20	
West Africa	20	
Indonesia	14	Highly collectivistic

Source: Hofstede.

For collectivists group harmony is more important than personal ambition. The group is the family, the extended family, the clan, the labor union, and the organization. The group is the source of one's identity. The group protects the individual who, in turn, is loyal to the group. At work, collectivists have a higher dependence on the organization and a stronger desire for non-financial rewards, such as physical conditions and benefits. The interplay of the collective organization and the collective family means that employees expect many special leaves for family events. Since relationships are essential to collectivists, then you must make a special effort to build friendships.[5]

Uncertainty orientation

Hofstede labeled this "uncertainty avoidance" and so many have reinterpreted this as having to do with risk. This is not about risk, but about a comfort with ambiguity. In high uncertainty-avoidance cultures, even when people hate their job, they will not switch. They prefer rules be set out and not broken. High uncertainty avoidance is found in Greece, Belgium, and many Latin nations.

Future orientation

Hofstede labeled this as Confucianism since the cultures with the strongest future orientation were all in East Asia, including China, Japan, and Korea. Future orientation is about delaying gratification for the future and includes characteristics, such as savings, thrift, perseverance, and persistence. The opposite of future orientation is an emphasis on the present or the past. In such cultures there is greater emphasis on tradition, social obligations, and more immediate satisfaction of material desires. Needs are gratified now, rather than in the distant future.

Time orientation

Cultures treat time quite differently, even though time is an absolute. At one extreme are those that see time as linear. Deadlines are firm and strict; people are punctual to meetings. Tasks are done one at a time. Germans and Americans tend to be in this group. At the other extreme are those that see time as elastic. Deadlines are flexible, meeting times are advisory and arriving at them 20–30 minutes after schedule is acceptable. Tasks can be done many at a time. Most cultures tend to cluster in this latter category including Latinos, the French (to some extent), and Indians. These cultural differences affect perceptions of the other. The linear time culture will tend to see the other culture as slow and inefficient, while the elastic time culture see the other as cold and rigid.

Communication orientation

This is often referred to by its two groupings: high- versus low-context communication. Communication orientation focuses on whether our communication – our messages – are specific and explicit. Low-context cultures listen more for what is said rather than how it is said. In high-context cultures, people consider the secondary, tonal, colorful, peripheral, *contextual* information in order to understand the communication. Northern Europeans and Americans are low-context cultures, while Southern Europeans, Asian, and Latinos are high-context cultures. Communication orientation helps us understand the use of e-mail as a communication medium. E-mail is more natural for low-context cultures, since they logically look at the information contained in the text of the message. High-context cultures, however, need the peripheral information, which is lacking in e-mail messages.

Destiny orientation

This refers to whether one believes that events are predetermined. This orientation is also labeled as *fatalism*. For example, whether a person is good or bad is seen by some as predetermined and one's behaviors cannot impact this destiny. Cause and effect are interpreted differently by those who see events and people's characteristics as predetermined. At the other extreme of the destiny orientation are those who believe in self-determination – the belief that their own actions determine outcomes.

Universalist orientation

A universalist believes that everyone should follow the rules, since the rules are meant to apply to everyone, to the universe. The other end of the spectrum are the *particularists* who care about the context of the rule infraction: perhaps it was a friend who broke the rule; perhaps she was sick the day before. There are significant differences between cultures on this dimension, as surveys repeatedly show. For example, in one study,[6] subjects were told of a case of an under-performing employee who had already accrued many years of excellent service to the company. Seventy-five percent of Americans and Canadians said to let her go, while only 20% of Singaporeans and Koreans did.

Information processing orientation

Relatively new research demonstrates that there are important differences in the way that Westerners and East Asians (Chinese, Japanese, Koreans) process information. This is described in a very readable book by Richard Nisbett.[7] Nisbett found that Asians tend to see relationships while Westerners see categories and taxonomies. For example, Japanese and American subjects were shown pictures of aquariums with fish of various sizes moving about, which were set in background of plants and rocks. The Japanese recalled 60% more background elements than the Americans and were twice as likely to recall relationships of background objects, such as the "little fish was above the pink rock." The Americans were more likely to recall the objects themselves.

Does culture matter?

In the software world one comes across the entire range of responses to the question: Does culture matter?

In one camp are those that were lucky and did not experience any cultural problems. The project went smoothly. This does happen sometimes. It is likely that there were some cross-cultural communication problems but they were attributed to something else – a difficult piece of code, an impossible deadline, an e-mail message getting lost.

- *Denial*: There is no cultural difference
- *Defense*: Other cultures are inferior
- *Minimization 1*: We're all human
- *Minimization 2*: The computer culture washes away differences
- *Acceptance*: Differences deserve respect
- *Adaptation*: Individuals and organization adapt to differences

---

Exhibit 9.1 Types of cultural acceptance and awareness.
Adapted from Bennett[8]

There are those who make this point in a more generalized sense: that software engineers are very much alike because of their common professional cultures, just like musicians and doctors. After all, software professionals all belong to the *computer subculture*. Software professionals, like engineers, place a high value on work and on achievement. American software guru Larry Constantine[9] maintains that the computer subculture is stronger than the national culture and that, for example, the Indian programmer will have more in common with an American programmer than, say, with an Indian government official. Constantine would be in the "Minimization 2" type of Exhibit 9.1.

We believe, however, that culture matters. In global projects the successful software engineers accept, respect, and adapt to cultural differences (these are represented by the last two types in Exhibit 9.1).

Examples of failure in cross-cultural communication

Examples of failures in communication across cultures are nearly endless. Nearly everyone who has worked across cultures has such stories. In perspective, these communication failures are not unique to cross national boundaries, since failures of communication occur between husband and wife, parent and child, employee and supervisor.

Given the importance of India as an offshore destination, these examples are about the Indian culture:[10]

- Indians are less likely to engage in small talk than most of their Western counterparts, as opposed to the British, who are said to "feel the terror of silence" in an elevator. The lack of small talk is interpreted as unfriendly. To make matters worse, some Indian providers train their employees not to ask personal questions to avoid a cross-cultural embarrassment such as: "Are you married?" No, "Why not."

- Indians tend to be too optimistic about times and schedules. If an Indian is asked how much travel time is needed to reach a certain destination, the answer is probably inaccurate and will not include the possibility of encountering traffic jams. This may stem from wanting to give a friendly impression. In work, this creates difficulties across cultures. Potential problems or delays are not considered. We heard a jest on this from an Indian cultural trainer: "When an Indian programmer says the work will be finished tomorrow, it only means it will not be ready today." With experience, one can anticipate this, by asking probing questions to find out which issues have been overlooked. Over time, and based on experience, you will able to calculate this "Indian factor".

- Indians, particularly from South India, have a perplexing habit, as they listen to you, of shaking the head in a manner that appears to be saying "no." This is labeled the Indian wiggle. But, it can have various meanings: "yes," or "I am listening," or "I agree," or "go on."

- Like many other cultures, Indians are reluctant to say "no." Indians may say yes when they mean "no" and they do not want to tell bad news. This is called a "wobbly" yes. For example: "Do you know ERP 4.5?" The Dutch engineer will reply "No"; an American engineer will reply "No, but I would be happy to learn"; and an Indian will reply with a "wobbly" yes. Again, probing further may unravel the wobbly yes. And, creating an atmosphere of trust and informal relationships will stimulate forthrightness.

- Making jokes can improve social interaction. However, what we consider funny might not be the case in another culture. When somebody makes an outstanding contribution to teamwork, the Dutch jokingly refer to such a person as "the best horse in the stable". A Dutch project leader used this phrase when complimenting a talented Indian programmer in the team. Although it was translated into English, the Indian programmer did not understand the positive meaning of the remark. He did understand however that he was being linked to an animal, and was deeply hurt.

Table 9.3 What the English really mean[11]

What the English say	What they mean	What is understood
I hear what you say	I disagree and do not wish to discuss it any further	He accepts my point of view
With the greatest respect	I think you are wrong (or a fool)	He is listening to me
Not bad	Good or very good	Poor or mediocre
Quite good	A bit disappointing	Quite good
Perhaps you would like to think about/I would suggest/It would be nice if	This is an order. Do it or be prepared to justify yourself	Think about the idea but do what you like
Oh by the way/incidentally	This is the primary purpose of our discussion	This is not very important
I was a bit disappointed that/It is a pity you	I am most upset and cross	It doesn't really matter
Very interesting	I don't agree/I don't believe you	They are impressed!
Could we consider some other options?	I don't like your idea	They have not yet decided
I will bear it in mind	I will do nothing about it	They will probably do it
Please think about that some more…	It is a bad idea. Don't do it	Good idea, keep developing it
I am sure it is my fault	It is your fault!	It was their fault
That is an original point of view	You must be crazy	They like my ideas
You must come to dinner sometime	Not an invitation, just being polite	I will receive an invitation shortly

Examples of failure in language

It is very difficult to understand what the British really mean, even for Americans. Given that the UK is the largest user of offshore services in Europe, some potential misunderstandings of their language usage is given in Table 9.3. Of course, these are humorous, but they contain a kernel of truth.

Americans, too, use deeply rooted linguistic code words. Most are not even aware of this special language. Within the rich vocabulary of business-speak, perhaps the two most important code words for cross-cultural software work are the words "challenge" and "issue." When an American says that the "interface of the G6 module is a challenge," she means that it is difficult. She may also mean, depending on context, that the difficulty may not make it worth doing and it should be dropped. When an American says that he "has an issue with the interface of the G6 module," then it means that he has a problem with it. He may even mean that he does not like it. Americans do not like to use the words *problem* or *difficult*, though they do sometimes use these words. They soften these words by speaking in code. Other cultures need to understand the true meaning of this code.

Johansson and colleagues[12] discovered an interesting case of language failure in their Swedish–Finnish student collaboration exercises. In one discussion a Finnish participant proposed that problems in the software development process not be discussed in detail. This would seem to be paradoxical since it is precisely the problems that should be discussed across distributed sites. However, it was discovered that the English word *problem* has two translations in Finnish: one is *tehtävä'* meaning a task to be solved; the other translation is *ongelma*, meaning trouble, with a suggestion that someone needs to be blamed for it. Thus, the Finnish did not report problems because he saw it as *ongelma* rather than *tehtävä'*.

In India, English is the national language of the educated class that participates in globalized software work, an inheritance from the colonial past of the British Empire. Many Indians do not speak English at home, however, but rather the local dialect, of which there are hundreds. Thus, their English, though fluent, has some gaps. Indians tend to speak English at fast but relatively even pace, with less intonation, and less stressed words. Yet, emphasis is critical to communication. Indians tend to take longer to explain things, which can be maddening to their impatient listeners from America or Northern Europe. It is useful to know the lexicon of words not to use. *Contractor*, a common term in the global business jargon, implying one who contracts for software, may also be understood in India to mean one who cleans toilets. Also, *vendor* is not always a respected term. So, use the term consultant instead.

Lu Ellen Schafer of Global Savvy shows the following e-mail message from India to illustrate Indian English:

> *Hi Joe:*
> *Let's prepone our conference call because there are a lot of things to discuss. Also, I have some good news for you. I found a rank holder to join our team and think he will be a fine addition. Having another person will help stop the cribbing I have been hearing!*
> *I do have a doubt about the project completion date.*
> *Regards,*
> *Vivek*

Each of the underlined words is difficult for non-Indians to decipher. Can you guess their meaning? *Prepone* is to move to an earlier time. *Rank holder* is one at the top of his/her class. *Cribbing* is complaining. *Have a doubt* suggests a tiny question rather than a doubt of the goal.

Schafer also tells the following story: your Mexican offshore partner informs you on the telephone that she will e-mail you the document now. But there are three words for "now" in Spanish: ahora, ahorita, and ya. Their practical meanings in this context, respectively, are: by the end of the day, within an hour, within minutes.[13]

Even names can be confusing. Often, Europeans or Americans, when reading exotic Indian or Chinese names, will not know if the name belongs to a man or a woman. It is somewhat embarrassing if a foreign female programmer is being referred to as 'Mr.' at the start of a project. And, many individuals like to be addressed using some name that is not their official name: an American named William may want to be called Bill; an Indian with a lengthy first name may ask to just use his last name in any communication. It is always a good habit to ask how someone wants to be addressed.

Technology and cultural differences

Much of our communications are conducted through the narrow pipelines of e-mail and telephone. It is useful to understand how these media help and hinder cross-cultural communications. Cultural and linguistic mistakes may be amplified because the communication cues are limited (see also the beginning of Chapter 8 on body language). Furthermore, because of the narrow channel, the communicator does not have the full arsenal of communications to soften a message, such as when discussing a difficult personal issue but smiling during the conversation. The "widest" channel is video-conferencing – but beware this technology across cultures! Due to different cultural orientations discussed earlier in this chapter, video-conferencing introduces another layer of sensitivities.

Every culture encounters communication problems over distance. Distance amplifies misperceptions. For example, Americans may be perceived as follows: rudely interrupting in video-conferencing meetings, since they tend to be more comfortable with quick, abrupt interactions; too informal over e-mail since they may not exchange pleasantries, use proper salutations, or may quickly sign with just the first name; impolite on e-mail, since relative to some cultures, they are blunt; and argumentative, since they tend to discuss and air disagreements and opinions in the open. In fact, in a study by Massey and colleagues[14] Americans had an easier time conveying opinions with distant partners, but a more difficult time dealing with convergence: in other words, getting to an agreement.

When communicating, the Dutch want to be clear and direct. If something is wrong, they will not hesitate to mention this. An Indian female programmer, who had recently started to work in a project for a Dutch client, was found one morning crying in front of her screen. She had just received an e-mail from the Dutch project leader, with a list of what she had done wrong. For her, it was a very impolite and unfriendly message. For the sender, it was just a number of topics which had to be corrected; it was definitely not meant personally. She would burst into tears on several more occasions over time, but she is now almost used to the way of communication in Holland. She still hates it.

Fluent English readers have an enormous advantage with the massive amounts of text that all must read: they can *skim* text quickly, hunting for key words or concepts;

quickly determining what is and what is not important in that long e-mail message. But many offshore partners cannot do this. They cannot skim and, thus, need more time to comprehend a long design document or a busy web screen.

Nevertheless, technology can also ease cross-cultural communications. First, for those who are non-fluent English speakers, it is easier to read and write than to speak. With e-mail they can read and write at their own (slower) pace than a native English speaker. This is very important to explain to fluent English speakers. For example, one American software architect we know has reduced his telephone interactions with his Chinese team because they had hinted to him that it is easier for them to read detailed specifications than to discuss issues via telephone. Second, e-mail overcomes the discomfort of understanding difficult accents over the telephone. We have often hear from Americans how difficult it is to understand some of their Indian colleagues even though they speak fluent English.

E-mail is preferred for other reasons. For example, in a Canadian–German software collaboration:[15]

> "… many of the German participants reported a reluctance to engage in argument over the telephone. When technical or methodological debates arose [… the German participants] reported that they preferred to have the time to formulate their position, write it down, check it, ensure that they were saying what they meant to say, and finally, send it off in an e-mail. While this addressed their discomfort, it introduced the potential for misunderstanding and stretched out the problem-solving exercise over an extended [back-and-forth via e-mail]."

And, by the way, this distributed project failed and management had to consolidate development in one location.

E-mail is also effective for helping to break hierarchies (power orientation) by encouraging people to choose direct communication (lateral communication) without going through the cumbersome hierarchy.

Social scientists are just beginning to understand the interactions of culture and technology. Sometimes they know the "what," but not the much of the "why." The diffusion of telework is indicative.[16] Telework is the extent to which employees work away from the office (either partially or in full). One would expect Americans to be among the more active teleworkers because of greater distances in America (they are). But the Dutch living in greater density have roughly equal telework rates. The French and Spaniards, on the other hand, have less than half the telework penetration of the Dutch.

Steps to improve cross-cultural communication

In this section we present a collection of recommendations, tactics, and tips for avoiding cross-cultural miscommunications.

Use of language across distance

- Try to speak and write in *International English*, the common core of British and American English, using simple sentences. For example, avoid phrasal verbs: instead of "I suggest we wrap up the project by June," say "I suggest we complete the project by June."
- Become aware of slang, idioms, and acronyms in your speech, and try to eliminate them. Sports metaphors need to be used carefully. Americans should not use baseball metaphors, British should not use cricket examples, and so on.
- Avoid contractions such as "can't".
- Avoid yes/no questions.
- Do not accessorize your sentences with synonyms (i.e. use the same word over and over).
- Be aware of words that have multiple or conflicting meanings across borders. "To table an issue" should never be used because it means the opposite to Brits and Americans.
- Explicitly state the response you expect. "Finish by Close-of-Business today," instead of "ASAP." "I will be arriving on July 23" instead of "I will be arriving soon."
- Keep e-mail messages short: one question, one response. Break up messages into short paragraphs and bullet points. Keep all sentences short.
- Use multiple channels to reduce miscommunication by repeating important messages redundantly: interlace the same message through e-mail and telephone and instant messaging (IM) and video-conferencing.
- Standardized, formalized terms should be used in e-mail messages whenever possible.
- Always remember the six 'R's:
  - Repeat. Go over it again
  - Reduce. Break it down
  - Rephrase. Be creative, use visuals
  - Reiterate. Emphasize the highlights
  - Review. Try to stimulate feedback
  - Recap. Summarize; in writing when necessary.

Inexpensive tactics for improving cross-cultural communication

- Buy and read a book about cultural differences to augment this chapter; then buy and read a book about the specific culture with which you are working.[17]
- Watch a good movie or read a good novel about the culture with which you are working. The successful movie "*Monsoon Wedding*" and the award-winning novel "*The God of Small Things*" are recommended for a better understanding of Indian culture.
- Talk to an expatriate at your office about the culture you work with. These people are bi-cultural. They can help you understand how those of other cultures perceive you.

Somewhat more expensive tactics for improving cross-cultural communication

The most important tactic for improving cross-cultural communication is training. These days, cultural training is easy to find. Numerous consultancies in every nation offer half-day or one-day cross-cultural training that can be tailored to the client's particular needs (e.g., how to work with Indians who are also engineers). Ideally, this training should take place before the first offshore project takes place.

Generally, it is the responsibility of the offshore service provider to train its own staff in dealing with the client's national country. The major Indian providers typically give their employees anywhere from 1 to 5 days of cultural training before they embark on an engagement abroad.

On the American side such training is too often neglected. We heard of a major New York firm that was asked by its Indian provider to provide the arriving Indian IT personnel with an orientation to the company, its culture, and the culture of New York City ("what is a bagel"). The company refused.

A few other tactics:

- Hire a bi-cultural person. Your hiring practices for offshore work should place a priority on hiring bi-cultural people.
- Learn the language. Of course, this is a longer-term investment. Even rudimentary knowledge of the language is helpful. At one medium-sized Russian software firm we visited, there were two full-time English teachers on the premises who gave continuous language instruction to individuals and to small groups.
- Conduct site visits. This is an opportunity to meet one's partners and get to know and understand their context and surroundings.

Case study Why the project was late: cultural miscommunication in an Indian–American collaboration

Lu Ellen Schafer*

Christina Salazar knew that Karnatec, an outsourcing company based in Bangalore, India, considered it a coup to land a contract with the leading Fortune 100 company based in California where Christina worked. As a senior engineering manager, she had selected Karnatec (an alias) over the well-known outsourcing heavyweights because she needed a nimble partner that would bring to bear all their top talent. Indeed, she knew these guys would bend over backward to make sure Christina's projects were a success. With fewer resources after a recent downsizing, this attitude was exactly what she needed since her team was working on a very aggressive schedule.

Christina understood there might be cultural differences with the offshore teams, but she was not overly concerned. She was good with people in general, and had a

* Schafer is at Global Savvy, USA

track record of reading them fairly accurately. The project managers in India assured her that the India teams were quite westernized. After all, they were part of the "Internet generation," and most of them had worked on global projects for several years.

In the Statement of Work, Christina defined two specific fixed cost projects, each with specific deliverables along an 8-month timeline. Vivek would manage the first project; Ashok would manage the second. Each project would require a team made up of 18 Karnatec engineers, plus two leads, working over 8 months. Christina and her two main project managers from California spent 4 days in Bangalore working out the details. Both teams committed to weekly calls and regular e-mail contact. Karnatec was clearly very excited about being involved in the project; Christina and her managers left Bangalore feeling relieved that they were all on the same page.

Both projects got off to a good start. The teams in Bangalore and California worked through the issues of time zone differences by being flexible. Sometimes the California team called India from their homes in the early hours of the day; sometimes the India team stayed late to accommodate California's time zone. The conference call adjustments typified the flexibility that Christina valued. In fact, Vivek further demonstrated his willingness to accommodate when he agreed to Christina's request to do all the migration testing. This freed up the California team to work on new features recently introduced by their competitor.

One month after they started the project, as agreed, Vivek's team had finished the testing. The quality met Christina's expectations, but, to her dismay, she discovered that Vivek's team had slipped the schedule of the initial project they had been working on. Christina, who was worried that she would now be in serious trouble, sent a pointed e-mail asking why Vivek had not informed her that he would be late on the initial project. Why wasn't this brought up in their weekly calls?

In the midst of trying to sort things out with Vivek, Christina and her team met with Ashok, the second project manager, who had come to California with his key lead.

Christina was aware that Ashok had a lot on his plate with Christina's rather complicated and dynamic project. Ashok's e-mails frequently stressed all the effort and long hours his team was putting in. Christina knew the required hours along with tight scheduling could make them feel overworked. Her own team in California was feeling stretched as well.

During their first meeting in California, Ashok laid out all that his team had been working on. Christina and her engineers could instantly see that the India team was significantly under-resourced given what they had to accomplish. One of Christina's engineers asked Ashok why he hadn't told Christina he needed more resources.

The Indian manager seemed exasperated. He said he *had* been telling them. Christina's team looked at each other in bewilderment. What was going on?

Christina began to wonder how they could have failed to communicate to this extent after all the clarity in the Statement of Work, their weekly conference calls, and the obvious intelligence of both her California team and the India team. In her mind, she ran back through the topics of the conference calls and e-mails but could not find clues that she had overlooked.

Lessons

In both of these incidents, Christina and the India team had stumbled upon a powerful but often unacknowledged cultural difference.

Edward Hall[18] first coined the terms *high and low context* to describe different communication orientations (a concept that was introduced earlier in this chapter). The interactions between the Americans and the Indians in this story illustrate the miscommunication that results from well-intentioned professionals from different cultural orientations. The US team, being from a *low-context* culture, had a propensity to be direct, to openly state their needs and expectations. The Indian-based Karnatec team, on the other hand, was from a *high-context* culture. Vivek, Ashok, and their teams, like many Indians and many outsource partners, had a tendency to be implicit, to assume that Christina and the US team would read between the lines. Their subtleties also evidenced deference shown to an important client. Besides, they knew Christina to be an intelligent, experienced manager who would not require that such things be spelled out.

In the first incident, Christina was surprised to find that Vivek's team had slipped the schedule on the initial project. From Vivek's *high-context* point of view, he was simply following Christina's priorities. Christina knew exactly how many engineering resources were devoted to her project, and exactly what they needed to accomplish in a given period of time. The implicit message, one which Vivek did not feel was necessary to reiterate as it was so obvious (to him and the Indian team), was that Christina had decided that the testing was the priority or she would not have taken him away from the other work to do it.

Christina, however, was operating from a very different premise. She assumed that if Vivek had accepted this new assignment, he could somehow get it done while completing the initial project as well. And if Vivek could not accomplish it all, Christina reasoned, he would say so.

The second incident stems from the same differences in *low* versus *high context*. Ashok felt he had been very clear with Christina about being under-resourced with all of the changes in Christina's projects. From his *high-context* perspective, all that was needed to alert Christina to the fact that his team was too stretched was to delineate what each engineer was working on. Since Christina knew how many engineers were working on the project in India, she should be able to clearly see Ashok's dilemma. From his perspective, Christina could do the math.

Christina, though, from a *low-context* position, expected Ashok to specifically state that he needed help if that was the case. When she received a list of accomplishments, she did not read between the lines to know that Ashok was actually sending a message of help.

Ultimately, everyone paid for the miscommunication. Despite the heroic efforts of both the California and Bangalore teams working to make up for some of the lost time, the two projects were late. Christina and her team lost trust in their outsource partner. Vivek and Ashok felt wrongly accused and under-appreciated for their efforts. Additionally, all the managers involved had the additional job of bolstering up their team's morale.

All the managers in this case were experienced, intelligent, and performed with the best of intentions. But with the clarity of hindsight, it is easy to see that much more communication was needed, especially enough to circumvent the roadblocks posed by a *high-context* team working with a *low-context* team across almost a dozen time zones and 10,000 miles. Given the complexity of *high-* and *low-context* communication, what can at first be seen as over-communication, is actually often just barely enough.

Case study In a Russian sauna with the Dutch manager

Julia Kotlarsky*

In 1997, Lizatec, a small Dutch software house, faced a dilemma: Where to get programmers? There were not enough good programmers in The Netherlands: Dutch programmers had a reputation for producing low-quality software and asking for high fees. The company started to look for opportunities to achieve better quality and reduce development costs. Lizatec started to think about offshoring, but to which country?

Lisette Breukink, one of three managers and co-owners of Lizatec, visited several companies in India to discuss the possibilities of outsourcing software work. However, this did not work out, mainly because of cultural differences. Lisette came back feeling that men in Indian companies could not accept the idea of working under the supervision of a woman. They just could not handle the fact that a woman would be their manager and give them instructions.

Lisette heard from some friends about Russian programmers and she convinced her Lizatec partners to try Russia. In late 1998, Lisette visited St. Petersburg, interviewed some Russian programmers, and a short time after that opened an offshore

* Kotlarsky is at the University of Warwick, UK

facility in St. Petersburg. By early 1999, 12 software engineers were working at Lizatec's Russian facility (and by 2004 there were 25).

Not everything went smoothly. There were many cultural adjustments that both sides had to make. Lisette related the three cultural vignettes described in this case.

Vignette 1

Dutch and Russian people have different perceptions of time. For example, the Dutch make appointments well in advance and work between 9 am and 5 pm, while the Russians are spontaneous and more flexible in working hours. However, you need to give them special treatment like inviting them to your house and cooking dinner for them while they are working, as Lisette did.

If a deadline is approaching and there is still plenty of work to do, what do Dutch software developers tend to do? At 5 pm they stop working and leave. What would Russian programmers do in this situation? When Russian developers from Lizatec were preparing the launch of a new product and a deadline was approaching, they told Lisette "Why don't we go to your place, you cook dinner, and we work and eat: it is more cozy to work at home." And this is what they did: Lisette went shopping for food and cooked dinner while the Russian programmers worked. Without knowing it, the Russian programmers took a small risk when they asked Lisette to cook for them, because Dutch women are usually not good cooks, in particular compared to Russian women. The Russian programmers were lucky because Lisette is not a typical Dutch woman and can cook nice meals.

Vignette 2

To motivate the Russian employees and keep them productive in the long run, you need to create a family-like environment at work. This is not typical for Dutch culture: Dutch people place strict boundaries between personal and organizational life, between home and work. The Dutch have formal relationships at work, while Russians need a home-like, friendly atmosphere in order to be motivated.

Lisette explained: "In the beginning the engineers made lunch by themselves, just dry sandwiches. They would come in early in the morning and by the end of the day they looked a little greyish and tired."

Then, Lisette hired a cook. She later explained: "I decided to hire a cook just to make a fresh salad and do some shopping, but she took her task very seriously and she started to cook three-course dinners in a corner of the office. So we ended up having a real dining room with a kitchen. And now every day we have a huge dinner at 2 o'clock." This is another cultural difference: in Holland people have dinner in the evening, while in Russia dinner is in the afternoon (2–3 pm), with a small supper in the evening.

Lisette emphasized that now programmers can eat all their meals at work: "If you are single, you don't need to do any shopping. Because you can come in and have breakfast (we have sandwiches for people who come in very early in the morning), then have dinner at 2 o'clock, and then leftovers are served for supper in the evening for people who want to work late. So there is always soup and salad. Now they can eat all day, and it is OK."

She continued: "The cook is now like a mother for everybody. If you are on a diet, she will cook for your diet. It is just her life – cooking for these people, like a family. […] And after this cooking, everybody started to look more healthy, because they ate real fresh vegetables, and fresh meat. This is my investment in everybody working well: everybody started to be more happy."

Vignette 3

In Dutch and Russian cultures the human body is perceived and treated in different ways, in particular the naked body. One might wonder: What does this have to do with offshore software development? It is indeed related: when the Russian developers (most of whom are men) go together with their Dutch manager (a woman) to a sauna ('banya' in Russian).

Russian programmers do not place boundaries between work and home. So, it was natural for Lizette's engineers to organize various cultural and social events so that their Dutch manager would learn about Russian culture. During the first year, Lisette would come to Russia once a month for about a week. "We went to the theater, paint-ball shooting, bowling, and to a music hall. Every visit somebody else organized such an evening together," she recalled.

One of the social events was to a summer house ("dacha") on the outskirts of St. Petersburg, to enjoy nature, have barbeques, and drinks for dinner, and, of course, use the "banya." This is where cultural differences became apparent. The Dutch are very practical, and treat their body as something functional: for them the sauna is associated with health and pleasure, and one cannot enjoy the sauna if he/she has something on. Thus, to enjoy the sauna fully, Dutch people take their clothes off and walk in naked. The mixed sauna (men and women together) is very common in Holland (as in Germany and in Scandinavia). For tourists and foreigners who live in Holland, there are signs by the entrance to a sauna saying that entry in swimming suits is forbidden.

However, this is not the case in Russia: Russians perceive the human body as essentially sexual. When Lisette asked her Russian employees how the Russian sauna worked – if they undress and just walk in – the Russian programmers were shocked. They told Lisette that in Russian culture the naked body is a sexual object: "if we see a naked body, we think about sex." Therefore, in Russia, in a mixed sauna, people put a towel around themselves or wear a swimming suit. Lisette was surprised: for her Dutch sensibility it sounded strange that one could be in a sauna and not be naked. "Strange, I don't think about sex when I am in the sauna, it is too hot," she said to the Russian developers. However, there was no choice but to respect Russian traditions and those of her employees. Lisette said to them: "Tell me what to do and I will do the same." This was the first time that Lisette sat in the sauna wrapped in a towel.

Case study Offshoring usability to India

Johan Versendaal,* Ramanathan Subramanian,** and Kaladhar Bapu[†]

Baan had been offshoring software development to India for many years but had never offshored usability. Usability is a bit different. Although the Indian engineers are respected and well-trained software developers, there is no such heritage in the domain of usability design (also called user interface design).

In 1997, the Dutch software firm Baan was a 1-billion dollar company[19] and one of the major global ERP companies competing with SAP and Oracle. Baan was one of the pioneer offshoring software firms, with operations in India since the late 1980s. By 1997, Baan had established large development and service centers in Mumbai and Hyderabad. Baan had established a mature offshore structure that ensured low-cost, highly standardized procedural software development. One of the characteristics of this maturity was that not only were Indian centers in charge of a great deal of *product development*; but that even some *product management* had been transferred to India.

Usability is a peculiar part of the software development life cycle in software product companies. In order to create high-quality interfaces the usability consultants visit customers, work closely with product management, and collaborate with product consultants and developers. The usability consultants influence the development process throughout the life cycle. Usability has no veto in the development process and must work through cooperation and influence.

Motivation for offshoring part of usability

Baan had been growing its product usability functions for some time, but these were all centered out of The Netherlands development unit. Baan was concerned that some principal product modules "owned" by the Indian centers would not include usability improvements. From a distance it was difficult for the Dutch usability consultants to interact with the Indian product consultants and developers. Furthermore, the Indian developers and product consultants did not have the expertise to include usability more explicitly in software development. Also, Baan-India thought it was important that there be close cooperation between local usability consultants and the local development group. Thus, the decision was made to start an Indian usability team.

* Versendaal is at Utrecht University, The Netherlands
** Subramanian is at Vanenburg IT Park, Hyderabad, India and Cordys R&D, India
[†]Bapu is at Cordys R&D, India and affiliated with the Indian Institute of Technology, Bombay

Establishing a multi-site team

In order to successfully launch the new team in India, a string of activities were planned over a year's time, starting with advertising and interviewing Indian candidates. Because Baan envisioned the Indian usability team working in close collaboration with the headquarters-based usability team, one of the Dutch usability consultants took part in all the interviews. However, the recruiting team could not find qualified engineers with backgrounds similar to the European usability consultants, who had a background in computer science, psychology, and task analysis. Nevertheless, five individuals were selected, fresh out of the university, who had studied visual communications, product design, and graphic design.

Knowledge transfer was carefully planned. Two Dutch usability consultants moved to India for 1 full year. Their main task was to help transfer domain knowledge and transfer process knowledge to ensure consistency in working methods and deliverables. They educated their Indian colleagues in the principles of usability design: task analysis, user study, cognitive science, and psychology.

During this period Baan needed to formalize its corporate-wide usability style guide. This effort became a joint project of the Indian usability team and their Dutch visitors during the transition year. This effort helped the Indian consultants to better understand the user interface design process. Simultaneously, the Dutch learned from their Indian colleagues to better understand graphic design principles and color theory.

During the transition year the Indian usability group was led by one of the Dutch usability consultants onsite. This was understood to be a temporary situation. Towards the end of the transition year, an Indian team-lead was found. By Spring 1999, the knowledge transfer and training was complete, the Dutch consultants returned home, and the Indian team started to report directly to the Indian-based product group.

Dealing with cultural differences

From earlier offshoring experience, Baan knew that they would encounter cultural differences. For example, Indian usability design professionals seek more freedom at work than their Dutch counterparts: they were not comfortable with the strict, punctual working style of a European company. As opposed to the more rigid working hours in Holland, the usability consultants in India started late each day, but ended up staying late and working even more hours.

The most difficult cultural difference was assertiveness. Culturally, the Dutch have an assertive, straightforward communication style. Usability consultants need to be especially assertive since they must be able to "fight themselves into" software development. The two visiting Dutch usability engineers recognized issues of assertiveness early in their stay in India – during the initial team-building sessions. In particular, when playing team games outdoors, the Dutch consultants would take the lead roles, while the Indians would accept secondary, supportive roles. In cultural terms, the Indian consultants accepted the hierarchy of the more senior and experienced Dutch – and did not question this. The team-building exercises sparked discussions about assertiveness. Consequently, during the transition year, the Dutch usability consultants emphasized assertiveness: coaching their Indian colleagues in attitude, first by jointly working on projects, later by evaluating work progress, and last by providing advice. By the end of the transition year, the Indian usability consultants had been thoroughly trained onsite to become more assertive.

Culture had another impact: on user interface design itself. When defining a screen icon, the palette of colors was chosen carefully to be sensitive to cultural, religious, and nationalist beliefs. Icons with hand gestures were rejected to avoid all possible cultural misinterpretations. The consultants' work together reinforced this rule, as they realized from their personal interactions that certain gestures, which were acceptable in one culture, were considered offensive to the other.

Lessons learned

Baan considered the transition successful in several respects. Knowledge transfer to the offshore team was complete; cultural adaptation was considered a success; the corporate-wide usability style guide, jointly developed by Dutch and Indian usability consultants, was fully applied by development; screen navigation developed in India had significantly improved; and icon design, also jointly developed, became a routine part of screen design.

All this was achieved at considerable cost: rotating two European consultants to India for an entire year. Unlike most offshoring efforts, establishing a usability team at Baan-India was not driven by cost reduction, but by product and process effectiveness goals. One of the conclusions from transition was that the organizational structure in India mirror the one at Baan headquarters in which usability was co-located with development. This would ensure clarity in communication and reduce ambiguity in roles and responsibilities.

There was one area, however, where missteps were made: giving more autonomy to the offshore team. Initially, the Indian consultants were allowed too little authority and too little design freedom. Often, work came predefined from The Netherlands, leaving little scope for creativity in India. It would have been better if both teams worked more in real collaboration that could lead to more mutual learning. Similarly Baan learned to give more autonomy on personnel issues and day-to-day management. Issues related to employee benefits, laws, policies, and practices required a deep local understanding and experience, and were best delegated to the Indian team manager.

Part III

Other stakeholders

10 Building software industries in developing nations

"… those countries in which science and technology is not applied as a guide to business, will fall behind and will be ever dependent on the development of others, for in today's society, those who use their knowledge and cleverness best, will be those who achieve advantage over others …"

José Mariá Castro Madriz, First President of Costa Rica,
Speech to Congress, September 15, 1844

"It is time to widen the scope of our participation in the knowledge economy from being mere isolated islands on the periphery of progress, to becoming an oasis of technology that can offer the prospect of economies of scale for those who venture to invest in our young available talent."

King Abdullah II of Jordan,
Speech to World Economic Forum in Davos, 2000
The King actively promotes the Jordanian software export sector

Software exports have become a cause for excitement in dozens of developing nations. These are nations that are all searching for the recipe to become the "next India." India, once known as a land of poverty, has now become an IT superpower. Major Western software companies, including Microsoft, Oracle, and SAP, are using Indian software centers to develop their products. India's independent firms compete with the top firms from the industrialized nations.

This excitement is understandable since many developing countries are primarily exporters of commodities, such as cotton or coffee. Other export sectors, such as textiles, handicrafts, assembly, or manufacturing, offer primarily low-skilled employment with low wages, often with "sweatshop" working conditions. And then in the 1990s, along comes software, a *high-skilled* industry suddenly within reach of developing nations. The current economies of software production, in particular its low-capital and high-labor intensity, is especially attractive for low-wage, labor-surplus economies.

Another allure is that the offshoring industry continues to grow year after year and that demand is forecast to continue rising for some years to come.[1] Nearly all observers agree that we are far from the saturation point. This growth creates new business opportunities for developing nations to attract employment and income-creating work. Governments, policy-makers, academics, and journalists are infected by this exciting potential. International organizations, such as the World Bank and UNCTAD, have recognized the economic and social impacts of a software export sector. National aid agencies from the US, Japan, Germany, Switzerland, and The Netherlands, among others, have all become interested in helping developing nations move forward by growing their software export industries.

The production of software is relatively environment friendly and does not depend on roads and harbors. Information, unlike products such as textiles or automobiles, can be transported quickly and cheaply through digital channels. Moreover, the required investments are modest, particularly when compared with industries, such as machine building or steel plants. Mastek, now a major Indian software company, began in a garage with only a few programmers and one fax.

There are about 150 developing nations in the world. This is a heterogeneous group, consisting of both low- and middle-income countries. Many developing countries have already begun software exports. These countries include some of the poorest nations in the world, such as Bangladesh and Nepal. Other nations are making preparations to enter this industry.

In Chapter 4 ("The Offshore Country Menu"), software-exporting nations were classified into three tiers. Based on the criteria of industry maturity, export revenues, and clustering, India and China are Tier-1 nations, along with many industrialized countries. They have large numbers of IT enterprises and their huge software exports are growing fast. Tier-2 nations are the emerging exporters, such as the Philippines, Malaysia, Pakistan, Mexico, or Brazil. They already have significant export volumes, although the number of exporters is usually less than 100. Tier-3 countries are still at an infant stage of exporting; examples are Cuba, Iran, Vietnam, and Indonesia. Their number of exporting firms is small, and so their volume of export work. This chapter is targeted at policy-makers, government officials, and other stakeholders from the Tier-2 and -3 software-exporting nations.

Choosing a national strategy

Once exporting software has caught their attention, nations in Tiers-2 and -3 are faced with a number of policy choices. Figure 10.1 depicts a simplified policy decision tree. While, it is unlikely that national policy-makers travel down this decision tree in a methodical fashion, it is a useful model to introduce the policy choices and trade-offs inherent in these choices.

Policy Decision No. 1 questions national investments in IT. Given the limited resources, is it worthwhile to focus those resources on the *production sector* (the sector that produces software)? Or, rather, should the country focus its resources on using IT for broader *information society* objectives, such as universal access, broadband access, e-government, or buying computers for schools? The debate about IT investments is most intense when it pertains to the poorest nations: Should the nation invest in a village's computing and connectivity when the village itself lacks drinking water,

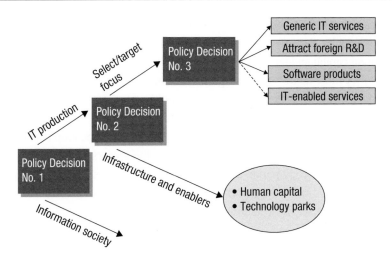

Figure 10.1 Policy decision tree for choosing a national strategy.

paved roads, and electrical power? As demonstrated in the Indian state of Andhra Pradesh and in research by the World Bank,[2] there are tangible results in using IT for reaching the poorest sectors of society by, among other, improving public administration, delivering remote health care services, distance education, and achieving gender equality. Dozens of developing countries have formulated national policies in these areas, labeled ICT (Information and Communication Technology) policies.

We believe that Policy Decision No. 1 is not exclusionary. Investment in IT production (software in this case) is needed to support broader information society objectives, otherwise the poorest nations will rely solely on foreign aid and foreign specialists to provide computing needs. Furthermore, as we will discuss in this chapter's next section, there is evidence that nations benefit broadly from a strong software sector, particularly if it is an exporting sector.

Policy Decision No. 2 is perhaps the more controversial question: How to encourage the software production sector? Should the software industry be encouraged by selecting a target vertical industry to focus national attention; or rather should the nation invest in foundational, infrastructure, and enabling elements? The first of these choices means that the nation, through its ministries and other governmental bodies, moves up the tree to Policy Decision No. 3.

It is our belief that governments should *not* select vertical industries. Governments tend to have a high failure rate when it comes to choosing on which technology to place a bet. Markets and technologies tend to move faster than government policy-makers. The best strategy for nations is not to choose a specific software industry strategy at all. The local companies will gravitate towards the more successful niches much faster than government policy-makers are able to react.

Rather, governments should focus on those areas in which they have clear strengths relative to market forces. These are areas that the market by itself cannot usually act successfully. These areas tend to be in infrastructure and other foundational areas. In particular, for software exports, there are two areas which have the greatest impact: human capital and technology parks. Investment in human capital will make the nation's industry more flexible and competitive in the long run. After all, competition in software is not just offering cheap labor; it is about low-cost *skilled* labor. Human capital investment includes programs in science and technology at national universities, as well IT-related programs in specialized technical schools and vocational schools.

The second policy, which we believe has the greatest impact, is to facilitate the creation of technology parks. Such parks make it easy for local or foreign firms to set up operations. A technology park is made up of one or more buildings in which there is a full range of services including reliable electrical power, high-speed lines, cabling, physical security, and so on. Foreign technology firms seeking new locations are especially drawn to these parks because they perceive them as being business friendly. And many nations also give tax incentives to those companies, domestic or foreign, operating in these technology parks.

These two foci, human capital and technology parks, are not the sole areas for governments to act. We note three other important areas that governments, together with industry associations, can affect change:

- In the short term, provide marketing assistance for those firms already exporting. As we describe in Chapter 11, building relationships with potential foreign clients require considerable efforts, especially if the provider is located thousands of kilometers away. Governments can offer marketing assistance by funding trade shows, leading delegations, and setting up match-making projects (to facilitate finding clients in key target markets).
- In the longer term, promote collaborative efforts between the government, software companies, financial institutions, universities, other educational programs, and business associations. Today these are often labeled public–private partnerships.
- Assist firms in attaining internationally recognized standards of quality. Such standards address a fundamental international marketing problem faced by providers from developing nations, namely, that clients see foreign suppliers as exotic and risky. The government ministry, together with the industry association, can set up training programs focused on quality certifications.

Which focus to select?

Nations that do travel to Policy Decision No. 3 have another difficult choice to make: deciding on a focus for their software export industry within a highly competitive environment. Some larger nations may choose to target all four foci depicted in Figure 10.1. Choosing all four may dilute any advantage of focus. At the same time, nations can choose more than one and there is some overlap between these foci.

Target No. 1: Generic IT services

This is the default choice: to be a commodity low-cost service provider. This choice does not appear to be glamorous since most Tier-2 and -3 countries compete primarily on low wages. Dozens of national industries, including thousands of firms, are offering commodity skills in programming (e.g. using Microsoft C++ or Java) with little national specialization and differentiation. Yet, just as there is room at the local fruit and vegetable market for many undifferentiated merchants selling very similar tomatoes side by side, there is room for a large global market in software programming services. This is not necessarily an incoherent path for small countries to select since most of them have no possibility of becoming the next Indian powerhouse. In fact, Indian firms face increasing competition from cheaper countries. By offering commodity skills, Tier-2 and -3 countries can find clients, build domain knowledge and later, perhaps specialize in service niches.

Ideally, those nations that focus on providing software services should find an identifiable niche. In such a case the nation builds a cluster of successful firms exporting services in an identifiable specialty niche. This is desirable since differentiated services are more profitable than those that compete solely on price. Furthermore, differentiated services reduce costs of marketing. While this is desirable, it is not easily attainable: we are not aware of such identifiable clusters in Tier-2 or -3 nations. Therefore, targeting a services niche for a national industry, while possible, is a formidable task.

Clusters of specialty may emerge when they are rooted in some other national competency. The model that policy-makers should strive for is to situate the industry close to strong domestic technology users in order to create synergies with these users.[3] For example, Costa Rica has chosen as its national strategy to build its "green" industries, to capitalize on its well-recognized environmental assets and knowledge. Therefore, software companies could situate themselves "in proximity" to these green firms and develop expertise in their special needs.

Target No. 2: Attract foreign R&D activities

The nation can attract foreign technology companies which will set up wholly-owned software development centers locally. This is quite desirable because the foreign firms provide capital and leading edge know-how. Additionally, foreign research and development (R&D) centers are the locus of highly skilled innovative activities in most nations. Most innovation activities in the new offshore destinations, including India and China, still take place within subsidiaries of foreign technology firms. For example, over the 25-year period of 1978–2003, foreign-owned software firms in India were authors of a total of 110,914 copyrights (the majority from IBM), whereas Indian software firms only had 208 copyrights.[4] Attracting foreign R&D is a desirable but difficult focus in which only a few countries have had any success: India, China, Israel, and a handful of others. Many developing nations are unlikely to be successful in this policy choice.

Target No. 3: Software products

Software products are attractive because, when successful, they can be far more profitable than services. Furthermore, software products are launched as a result of greater innovation intensity, which is a characteristic that nations strive to achieve. Unfortunately, national success in exporting software products is quite difficult. The basis for competing in software products is deep domain knowledge (in business, science, or some other domain) coupled with strong managerial and marketing skills. Most of the nations that began competing in software lack these skills inside their industry. Furthermore, software packages require capital, since the production investment comes upfront. International marketing of software products is usually more difficult and costly than the marketing of services. Low labor cost is less important.

With the exception of Israel and Ireland, none of the new offshore nations has yet made a mark in this sector. Even India has had very limited success, and that, only after decades of intensive knowledge transfer from abroad;[5] and after years of urging by observers to diversify from low-end services. There are a handful of examples of *single* firms from developing nations successfully exporting software products. Signum from Ecuador succeeded in selling its Spanish language checker to Microsoft which incorporated the product in its Office suite. STA, a firm in the Philippines, developed and sold a Year 2000 conversion tool abroad. Such examples, however, do not represent more than the success of a single, individual software company, not of a national industry. We have also found that some companies from developing nations attempt to market software products that are inappropriate for the markets in industrialized nations. Their naïveté about client needs is another reason to stay away from software products.

Target No. 4: IT-enabled services

Smaller developing nations may even choose to migrate away from software or "skip" IT altogether and concentrate their national resources in specialized areas within the IT-enabled services (ITES) sector. An example is the Philippines, which has seen success in exporting IT services, but has been far more successful in exporting IT-enabled services.

Sri Lanka may also be shifting its attention.[6] A window of relative peace from its civil war created a new focus on opportunities for economic growth and exporting, and, in particular, exporting IT-enabled services. The IT-enabled service firms in Colombo have traditionally focused on a few simple operations, such as data entry, but there is interest in expanding into new areas at both the lower and higher end, including accounting, legal, and engineering services. Engineering design drafting of the Dubai airport was undertaken in Colombo. The first offshore call center was commissioned in 2002 and uses voice over IP.[7] Indian IT-enabled service providers saw opportunities in Sri Lanka: for example, WNS Global Services began call center and transaction-processing operations. In the meantime, in 2002, the government embarked on an initiative to formulate a comprehensive national ICT development strategy called "e-Sri Lanka: ICT development road map." In addition to the usual national ICT components, such as connectivity, legal reforms, and human resource (HR) development, the plan included some actions aimed at the IT-enabled services export industry.

ITES: the policy choice of skipping software altogether

For some years now, the offshoring of IT-enabled services has grown faster than the growth of IT offshoring. It is "the next big thing" in offshoring and many Tier-2 and -3 countries have focused their attention on this area. As we described in Chapter 1, offshoring ITES is closely tied to offshoring of IT in a number of ways: the providers are often identical, the clients are often the same, the facilities are often comparable, and the managerial know-how is somewhat similar.

ITES also has some important advantages vis-à-vis IT: some IT-enabled service areas have low-entry barriers in terms of skills, scale, technology, managerial capabilities, or domain knowledge. This makes IT-enabled services more accessible to the least developed nations. It is a sector in which even some African countries are active. For example, Ghana, Mauritius, Morocco, Senegal, and Tunisia are hosting call centers, some of which are French speaking and are targeting France.

Nations considering IT-enabled services have many specialties to choose from. These specialties fall into one of the following three categories: customer interaction services, back-office operation services, and data and content integration.

Customer interaction services

These are front-office activities, where employees are in direct contact with customers. These services are mostly voice based, with call centers and helpdesks being the most common. Many nations can offer language-based services – provided that reliable, fast, and cheap telecommunications is available. Many thousands of call-center jobs have been transferred to English-speaking countries, such as India or the Philippines. Mexican call centers are serving the Spanish-speaking customers of American companies. China is offering Japanese-speaking staff to service Japanese firms. Other nations have been slower to grow their exporting of IT-enabled services. South Africa's call-center industry suffered from underexposure in the international marketplace. The nation has some clear advantages for European users, such as cultural similarities, common time zones, and political stability. The industry awoke to the opportunities as evidenced by the aggressive marketing overseas by the Western Cape Province, home to Cape Town. Governments can help grow customer interaction services by facilitating training in languages, accent neutralization, accent comprehension, telephone etiquette, listening skills, and problem/objection-handling skills.

Back-office operation services

Some of the back-office work has very low-entry barriers in terms of skills and can be done in many developing countries. One of these services is the handling of reservations. This is a high-volume, low-value type of work where competition is based on factory-like approaches and low wages. Thus, American hotel chains have turned to Jamaica and British Airways, in 1996, turned to India.

Other services require some specific knowledge. Claims processing for medical insurance companies allow doctors and hospitals to pass on all relevant records (via scanned images) to the offshore staff which handles all the paperwork. This work requires knowledge about the details of medical insurance. Understanding medical words and phrases is also required for medical transcriptions. The conversion of dictated voice recordings of doctors to paper is a niche market for specialized offshore service providers.

Document imaging also allows offshore locations to perform finance and accounting services, such as processing accounts payable, accounts receivable, financial reporting, tax consulting, or internal audit services. Specific domain knowledge is required for these functions, and some countries possess these skills. The Philippines has a large supply of accountants trained in US accounting standards. US-based Procter & Gamble performs its accounting services in the Philippines, where 650 employees complete the corporation's tax returns for its global operations.

Data and content integration

Data entry is required when data, stored on paper, needs to be converted into digital form, and scanning is not an option. In those circumstances, manual data entry is a solution. Data entry requires the lowest level of computer literacy and does not require strong language or functional skills. It also requires very little interaction between the customer and the offshore provider. It is therefore a service many Tier-2 and -3 countries can offer. An example of data entry is the official Dutch dictionary, which consists of 40 volumes. When it required migration to digital format, the books were sent to India, and keyed in twice by different groups of typists. The two data sets were automatically compared to find any differences and then corrected.

Digitizing is the process of converting texts, images, video, or records into digitized forms. The advantages are long-term preservation of the documents and easy access to the information. It is a labor-intensive activity, and ideally suited to be done in a low-cost country. The imaging is sometimes done on the clients' premises. In other cases, the original records, or copies, are shipped offshore. Anglo-Dutch publisher Reed Elsevier is digitizing all its scientific publications, printed over centuries, at SPI in the Philippines. A Swiss newspaper is digitizing its historical publications at Dakor, in North Korea.

Geographic information systems (GIS) are increasingly used by governments, telecommunications, and cable television companies. Much of the world's geographic data still resides on paper sources and needs to be converted. A large proportion of the cost of GIS is in data collection, interpretation, and conversion, making offshoring a viable option. GIS providers can also be found in several Tier-2 and -3 countries. For example, one of the leaders in this niche in India is Rolta with more than 2000 employees. Rolta is also moving to higher value activities by expanding into technical services, such as CAD modeling, plant design, and mechanical design.

Another area of IT-enabled services growth is that of animation services. Animation (e.g. cartoons) is very labor intensive to produce, even when computers are used, and is therefore ideal for offshoring to low-cost nations. These services include the production of a scenario, a storyboard, the drawing of character models and backgrounds; as well as the intermediate process of layout creation, animation celluloid, scanning, coloring, and composition. Countries such as India and the Philippines are offering animation services. Interestingly, North Korea, with very low labor costs, has targeted animation. SEK Studio in Pyongyang is one of the largest animation studios in the world with an artistic staff of 1600. It produces for French, Italian and Spanish film and TV companies.

Why developing nations should invest in building a software export industry

Software exports demonstrate the advantages of globalization without most of its negatives. The industry is a clean, non-polluting industry employing people using their brains rather than their muscles. Software is also a *general-purpose technology* that is used across all industries and therefore has positive impacts across the economy. Policymakers in developing nations recognize that such high-tech exports are more valuable to their economies than exports of coffee, or minerals, or assembled goods, or tourism.

In this section we introduce some of the advantages of a software export industry, namely the creation of jobs, revenue generation, improvements within organizations, and other positive impacts on society.[8] The latter benefits are spillover effects, or as economists refer to them, *externalities*. Spillovers are the benefits outside the software-exporting sector itself – to other businesses and to other aspects of society. In summary, the software sector (domestic and export oriented) is important to national well-being because it has multiplier effects: it is a general-purpose technology and because it creates spillovers.

Many of the examples in this section are from India, where the impacts of the software export industry have been relatively most dramatic.

Job creation

Unemployment is a major and persistent problem for all developing countries with insufficient opportunities for educated youth. Sadly, acquiring a university degree is no guarantee of a job. Due to unemployment, many countries experience labor migration, resulting in a brain drain. To address this dire situation, the promise of the software export sector is job creation for the nation's educated citizens. Nations have been looking with envy at India, where the software export sector offers direct employment to more than 260,000 people in 2004 and ancillary jobs are at least double that number. In the Tier-2 and -3 countries, the amount of software jobs is still modest, but growing. For example, the Vietnamese IT sector, which we described in Chapter 4, employed about 10,000 software engineers in 2003 but was growing at a 30% rate. Even more enticing for developing nations, the promise of job creation in ITES is even greater.[9]

Importantly, the IT jobs created are higher skilled than those in most other sectors of an economy. For example, more than 83% of the employees in Brazilian software companies completed a secondary education, compared with 39% national average. The number of employees in software organizations with a university degree is twice as high as the national average.[10] By providing interesting and rewarding opportunities to the educated, the software export industry can retard, or stop, or even reverse the emigration of highly skilled labor. There are several dramatic examples of this: in Ireland in the 1990s, and in India and China in the early 2000s. Indians and Chinese working in Silicon Valley are returning home to set up businesses. The brain drain gives way to "brain circulation" and becomes a "brain gain" for the developing nation.

Revenue generation

Exporting software is a source of foreign currency revenues, and India is again a spectacular example. We can mark the beginning of the Indian software export sector in 1973, when Tata Consultancy Services (TCS) began exporting data services to Burroughs. By 1983, the Indian software exports were estimated at a modest 18.2 million USD.[11] Some two decades later, in 2004, the value of the Indian exports increased to 12.5 billion USD. This seven hundredfold growth in 30 years is already responsible for 20% of India's exports[12] and projected to become India's single largest export industry within several years. An important part of the earnings, estimated at 55%, stays inside India.[13] The exports earnings of employees within the IT sector are also high. A software programmer can generate more than 10 times foreign currency than an employee working in the garments industry.

Improvements to the national business culture

One of the pleasing spillovers that are evident in India is the *demonstration effect* of its successful software export industry. Firms outside the software industry are learning and imitating the new business practices of the successful software firms across many dimensions: in working conditions, professionalization of HR practices, in new organizational structures, and in embrace of international standards.

Within India's software industry the larger companies offer exemplary working conditions in modern, air-conditioned offices. Employees are pampered with benefits from meals to company-provided transportation. In a traditional country such as India, distinctions based on religion, sex, or caste are less important in the merit-based software industry than elsewhere in the country. Indian software organizations tend to be flatter with more employee participative decision-making. Additionally, professional HR policies are instituted to hire and train labor. Staff retention policies were introduced. Employee incentive plans, such as Employee Stock Option Plan (ESOP), previously unheard of, have become common. Some companies are adopting the P-CMM quality model (People Capability Maturity Model), a framework for enhancing the competencies of their staff.

Companies are professionalizing in other respects, by adhering to a myriad of international standards: international accounting standards (GAAP), professional corporate governance, professional marketing standards, and quality standards such as ISO 9000, the CMM, or Six Sigma. Indian software firms practice some of these standards with greater success than do most Western organizations, and with the rise of IT-enabled services, Indian firms have rushed to embrace its new standard COPC.

Most interesting is the impact on the "pop" business culture. A new class of heroes is celebrated and imitated: they are the software entrepreneurs who have become rich and created national wealth in the process. Azim Premji, the founder of Indian giant Wipro, appears on the Forbes list of the 50 richest people on earth and is nicknamed the "Indian Bill Gates." He is an example of the growth of a technocratic innovation and entrepreneurship model previously treated with some suspicion in India's statist culture. Such new heroes are also appearing elsewhere in East Asia.

Other economic and social impacts

Software export success has a positive effect on the domestic software sector. Working on offshore projects is a form of *knowledge transfer* from the wealthy nations to the developing world: the technical and domain knowledge gained through working for foreign clients can be re-channeled and used for domestic projects.

The software export industry also induces investment in infrastructure. It spurs investments to re-haul developing nations' antiquated communications infrastructure, which benefits other economic sectors. In addition, the software industry creates demand in various services, such as transport, construction, accounting, hospitality, and legal. Demand for software skills generates investments in general education, higher education, and specialized training institutes.

The wealth created by the software industry can make some locations more attractive by spurring arts and entertainment to flourish – these are improvements in the quality of life. Some of the industry wealth is also channeled for social philanthropy. Indian examples are the Infosys Foundation, which is active in areas such as learning, rural development and health care; TCS has an Adult Literacy Program. The founders of Baan set up a school near Hyderabad, where the company had a development center.

Impact on the digital divide

In poor societies millions of citizens are further marginalized by having no access to computing. Unfortunately, a successful software export industry will not automatically diminish this digital divide. The impact of the export industry is often limited to a small sector of the economy. In India most software development takes place in only a handful of metropolitan areas, not touching the country's vast underclass. There is a danger that an export sector will be an enclave in the economy with limited forward and backward linkages. The discontent with the (digital) divide was exemplified in the surprising loss of India's ruling party in the 2004 national elections. In spite of all the wealth created by the new digital elite, the majority rejected the *status quo*.

Principal success factors

Given the numerous advantages of a booming software export industry, it is understandable that government officials and industry leaders in developing countries have been trying to create their own successful industries and to emulate the Indian success story. What, then, is the secret recipe? What has brought about the success of some countries in software exports? And, what factors are likely to foretell success in others?

Simply copying the success of others is not possible, but policy-makers can learn from others' experiences and assess their own strengths and weaknesses. When one examines the range of nations exporting software, eight principal factors surface which explain national success in this industry. These eight factors make up the "Oval Model", first described by Carmel,[14] and labeled *oval* because of the shape of the national boundary depicted in Figure 10.2. This model can be used in order to look back and

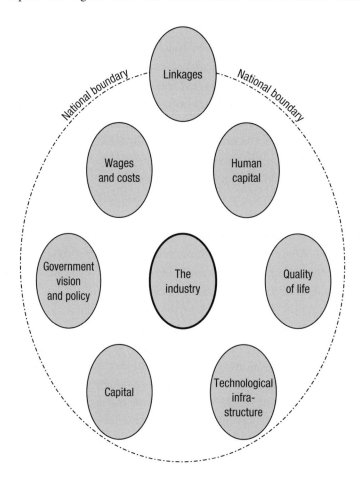

Figure 10.2 The "Oval Model" depicting success factors for a national software export industry.

explain the success of those nations that have already made achievements in exporting software. It can also be used as a framework for prescriptive policies and strategies.

The eight factors in the Oval Model are not independent of one another and tend to influence and interact with each other. Perhaps most important to note is that not all the Oval Model factors need to be present to achieve some success; in fact in India, during its industry's formative period of growth, several of the factors were absent or weak.

Government vision and policy

Governments in dozens of nations as diverse as Costa Rica, Iran, Indonesia, Bangladesh, Vietnam, and China are taking concrete policy steps to promote their software export industry. In Jordan, King Abdullah was personally involved in the 1999 launching of national plans for its software industry setting very ambitious goals.[15] Thus, government is often a major actor and a major success factor. It can play either a proactive kingpin role, or an enabling role, or both. Of course, the government has influence on every one of the other factors in the Oval Model.

National visions for software, or computing, have changed over the years. In the 1980s the fashion was to protect the industry as much as possible from international competition, rather than opening it up. Brazil protected its computer industry for a decade, as did India. In India these *swadeshi*, or self-reliance policies, had some positive impacts on the largely homegrown nature of the industry, since foreign IT companies were kept away (IBM even left the country). On the other hand, the severe restrictions on inward foreign investment and the high-import tariffs on equipment had a negative effect on the industry's growth. Later, the economic reforms in 1985 and 1991 coincided with the dramatic growth of the software export sector. The form and content of state intervention changed into that of a facilitator of private sector initiatives. The Indian government's role is primarily about providing an enabling environment through supportive regulations, incentives and strategic investments, and promotional programs.

Government actions are most effective in areas where markets are "inefficient," such as education. Investing in human capital, described further below, is government's key long-term role for the software export industry. Government can also play a facilitator role by encouraging ties between universities and the IT industry. In addition, governments can take a customer-centered approach by reducing the bureaucratic hurdles required to start a local company or in attracting a foreign company. Governments can establish a "one-stop service" for international customers, or give tax incentives favorable for software companies.

The government can build and guide national infrastructure in two vital areas: technology parks and telecommunications. For example, technocrats in the Indian government had the foresight way back in 1986 to make a set of recommendations captured in a report titled 'Software Policy on Computer Software Exports, Software Development and Training.' The report contained a set of actions that were targeted specifically at the software industry, including the creation of Software Technology Parks (STPs) that were to offer reliable electric power and adequate international telecommunications links. By 1990, three STPs were established in Pune, Bangalore, and Bhubaneshwar, growing to 18 STPs by the early 2000s. The STP companies account for 68% of India's software exports.[16] Spurred by demand of its already-booming software industry, the Indian government woke up in 1999 to the need to reform its inefficient monopoly telecommunications system. A large number of private providers were allowed to enter the business. Results were dramatic. In the major cities, the quality and the cost levels of the telecommunication networks are now approaching the levels of industrialized nations.

Human capital

Human capital is a nation's key resource. The term *human capital* is used to connote the knowledge and capabilities of the nation's workforce. Nations with poor educational systems and lacking in effective organizations and institutions (in which its citizens learn to work) may have many workers, but little human capital. The nations that invested in their citizens many years ago, by building strong universities, polytechnics, and vocational schools are now reaping the benefits from this investment. We see the human capital success factor in software exports as having three pillars: science and technology human capital, organizational human capital, and linguistic human capital. We discuss each of these below.

The workforce in the software export industry come from a variety of disciplines. Some are trained specifically in IT (in computer science, software engineering, and other related disciplines), but many come from engineering, physics, mathematics, or even chemistry. Therefore, it is the broad expanse of human capital in science and technology that is a nation's success factor in software exports. Table 10.1 displays some estimates of the size of the IT-related workforce pipeline. Interestingly, in the offshore era a number of small nations with weak educational systems have tried to import human capital. For example, some Jamaican firms recruited programmers from India to conduct export-oriented projects, with limited success: the well-qualified Indian recruits could not be retained and left after a short period for the US.[17]

Table 10.1 A sample of the workforce pipeline of IT graduates[26]

India	99,000 (2004)[18]
China	70,000 (2002)[19]
Philippines	17,000 (2001)[20]
Iran	15,000 (2002)[21]
Brazil	14,000 (1999)[22]
Indonesia	5000 (2002)[23]
Vietnam	3500 (2003)[24]
Bangladesh	3000 (2002)[25]

The second pillar of human capital is organizational capital. This is the skill of managers to execute and implement coupled with the skill of staff to operate effectively within teams and organizations. Organizational capital is acquired slowly over many years through relevant experience. In part it can be learned. The part that is learned is called managerial skills and can be learned in management and administration schools. The boom in business schools around the world will improve organizational capital in many countries. In the software industry, managerial experience is most acutely missing at the mid-level for project managers. This is even a challenge for India, which possesses many people with technical skills, but far fewer experienced project managers.

Finally, we come to the third pillar of human capital, linguistic capital, the ability to speak the language of your client. The dominance of the English language places software producers in non-English-speaking countries at a disadvantage in the global market. As we described in Chapter 3, English language skills are one of the criteria used by clients to select an offshore destination. Countries with an historical relationship to the UK or the US, such as India, Pakistan, Bangladesh, or the Philippines, enjoy advantages in this regard. Others, such as China, Indonesia, or Vietnam, have language difficulties and lose opportunities as a result. These nations are investing in language skills for their younger citizens. In the shorter term, software companies are trying to fill the gaps by investing, by themselves, in internal language training for their staff.

Wages and costs

Clearly, it is low wages that are a key factor in offshore nations' success. This assertion is exemplified by the shift of work away from offshore nations that have become relatively expensive. The key "offshore" nations of the 1990s, Ireland, and Israel, can no longer compete on costs. In a different vein, is the current shift of work from India. Even in India, where a heated market is pushing up wages, pushing work to lower-wage nations, such as Vietnam and China. More than a dozen of the largest Indian IT vendors have set up development centers in China.

In the long run, the only way for a nation to escape this cost-driven spiral is to differentiate its IT work. Clients will return because of factors other than costs, such as knowledge, specialization, and excellent service. In the short term, in order to compensate for rising wages, other cost-related factors can be manipulated in order to lower the overall costs for the offshore client. Such factors include taxes, cost of office space, cost of infrastructure, and cost of training. Governments have the power to help reduce these costs to compensate, at least in part, for rising software wages.

The industry

The software industry is a collection of individual firms that have certain characteristics that they share as a group. These characteristics can make for a successful-exporting industry. We call them the three "Cs": concentrate, compete, and cooperate.

A *concentration* of firms is called a cluster and is quite familiar from Silicon Valley. In a cluster software organizations are close to one another, perhaps in a technology park, perhaps on the same edge of a metropolitan area, near universities or research institutes. Successful software firms are often found in regions where many other software companies are also located. If there are universities close by, then these are usually the source of skilled employees. Some clusters are government policy initiatives, such as the Multimedia Super Corridor in Malaysia, but other successful clusters seem to arise organically, without much government action, such as Silicon Valley.

Cluster effects have a positive benefit on each individual firm in the cluster, independent of any other strength or weakness of each firm. The labor mobility within a cluster enhances the exchange of (tacit) knowledge. Clustering creates competition, which spurs companies to innovate, to increase productivity, and to differentiate. It also fosters cooperation, which spurs growth and facilitates the sharing of knowledge. For example, in China, the Nanjing Software Export alliance was set up by local companies and has been helpful to increase exports.[27]

Many of the dynamic clusters are now in Asia. Major Chinese software clusters are located in Beijing, Shanghai, Shenzhen, and Nanjing, mostly in technology parks. Such large clusters function as a one-stop shopping location for international customers. The Indian software industry started in Bangalore, and other clusters developed later in Mumbai (e.g. Santa Cruz Export Processing Zone), Hyderabad, Chennai, and Delhi-Gurgaon. A cluster can acquire a strong reputation in and of itself, and becomes a geographic brand and gives international credibility, as in the examples of Hyderabad ("Cyberabad") and Bangalore ("the Indian Silicon Valley"). Bangalore is now home to more than 1400 software organizations and employs 150,000 software engineers, which is more than Silicon Valley with 120,000.[28]

A nation's software export industry cannot succeed without a critical mass of companies. A critical mass may be 10, or 50, or 100 firms. This number will vary by nation. We also contend that at least a handful of companies be of some significant size. In Bangladesh or Nepal, companies of 100 people are considered large: this implies that their software export industries will not grow very much. Small firms cannot win large contracts. In terms of sheer size, for example, the top three Indian providers, all are at 30,000–40,000 employees. These Indian enterprises act as a nucleus around which smaller software firms pollinate. Ex-employees of these companies have become active entrepreneurs and set up new software companies.

Companies can *cooperate* in a number of ways, but most importantly, at the early stages of a software-exporting industry, is to cooperate in an effective national association. This is the organization that promotes the nation's industry abroad, which is a topic that we will return to in Chapter 11. A national software association can also provide services back to its member firms, such as supplying information about local and foreign software markets. NASSCOM has been very successful in branding India as a software destination and other industry associations are attempting to emulate this success. The Indian association has also been instrumental in lobbying the Indian government for favorable tax and regulatory changes. Many developing countries do not yet have effective industry associations.

Capital

A software export industry needs capital in order to grow. Most companies in developing nations grow using their own capital, or stated differently, they are self-financed. But this restricts their ability to grow and prosper. Outside capital for growth can come from either foreign or domestic sources. Domestic sources include government funds, bank loans, venture capital, investment capital, and equity offerings. Sources of foreign capital are foreign loans, venture capital, investment capital (FDI), foreign equity offerings, and foreign aid.

The large Indian providers have no more problems raising capital. They are traded at stock exchanges, both in India and abroad (e.g. the American NASDAQ). But smaller and younger offshore firms do not have such access to capital. Only the Israeli software firms, during the 1990s, were able to fund *young* firms, rather than established firms, from foreign risk and equity sources. However, for most software firms in developing countries, the difficulty of obtaining financing is a major obstacle. Unless there is substantial collateral to secure loans, software firms have little access to funds from conventional financial institutions. In sum, software firms do not have adequate access to capital and must rely on their own working capital.

National governments can play a role in a number of ways: they can provide financial assistance through grants and loans, they can guarantee loans, they can seed risk funds (venture funds). They can also fund international marketing efforts. Separately, they can build technology parks and subsidize rents for promising software companies. Some international organizations provide funding that targets high-tech growth. The World Bank invests in offshore service providers through its International Finance Corporation (IFC) program and has provided equity investment to such companies as Systems Ltd. from Pakistan, Glass Egg from Vietnam, and IT Worx from Egypt. The Netherlands Development Finance Corporation invested in Eastern Software Systems from India.

Technological infrastructure

In much of the developing world electrical stoppages are a daily occurrence, requiring generators. In quite a few developing countries the telecommunication structure is still outdated and the costs of communications remain high.

Of course, small software firms can tough it out and operate with poor infrastructure, but the industry will never thrive under such conditions. Firms need reliable connectivity and it must be affordable compared to international levels. For example, in the case of call centers, telecommunication costs can represent up to 40% of the total costs. Access to fiber-optic links is important for countries that seek to attract call-center activities.

Most governments have already acted to upgrade their telecommunications, though some have not moved fast enough or moved to aid the software industry specifically. In situations where the infrastructure is absent on a national basis, technology parks or high-tech office centers are alternatives. Technology parks can now be found in developing countries around the world. An example is Mauritius, a small island nation in the Indian Ocean. It is creating the new Ebene CyberCity, a combination of commercial, residential and enterprise infrastructure. The heart of the new city is a 12-storey Cyber Tower. With this center, the island is successfully attracting foreign IT investments. Infosys of India is in the process of setting up a disaster recovery center on the island.

Linkages

Linkages (also known as bonds, or connections, or ties) are a vital part of doing international business. They facilitate the early getting-to-know you stages and, equally important, they ease the day-to-day problems of communication and coordination. Linkages emerge between individuals, between companies, and between nations due to geographic, cultural, linguistic, or ethnic connections. The effective use of linkages is one of the most important success factors for developing a software export industry. We will also return to the topic of linkages in the next chapter.

Linguistic linkages are illustrated by the success of India which is partly due to English fluency in this former British colony. African francophone countries, such as Morocco and Tunisia, are working for French customers. Companies from Arabic-speaking Egypt and Jordan are working for clients in Saudi Arabia. Spanish-speaking projects go to Costa Rica, Argentina, and Mexico. South Africa, where a very old version of Dutch is spoken, is targeting Dutch clients for its call centers.

Geographic linkages can be used in the case of nearshore outsourcing. Mexico and Canada have an edge over India in terms of their proximity to the USA. Eastern Europe is a nearshore source for Germany. Indonesia or the Philippines are logical destinations for Australia.

Diaspora linkages have been a powerful success factor. These linkages are a source for knowledge and technology diffusion, they transfer capital and they create business networks with their home countries. The Indians have made masterful use of these linkages. The non-resident Indians (NRIs) came to the USA for advanced education (such as an MBA or engineering degree), stayed and rose to influential positions in high-tech companies. These NRIs became the biggest champions of offshoring. They were used by Indian IT providers to create contacts, to gain initial sales contacts, or acted as marketing agents. About 80% of the Chinese professionals in Silicon Valley came first to do either a master's degree or PhD., typically in fields like electrical engineering or computer science. Together with the American-Born Chinese (ABCs), the new waves of educated Chinese immigrants are establishing business links back to China.[29]

Quality of life

The quality of life in a location helps attract foreign clients. Equally important, it will also help keep the best employees, the talent, from moving away or emigrating. The American professor Richard Florida calls these people the "creative class" and argues that locations need to have quality of life in order to draw them or keep them.[30] Locations with high measures of quality of life have several common characteristics: quality of place (natural, recreational, and lifestyle amenities), an abundant supply of labor, and high levels of environmental quality.

In India, Bangalore has always been viewed as an attractive city. It used to be known as the "Pensioner's Paradise", because ex-civil servants preferred to settle in its relatively temperate climate. This relatively less hectic metropolis has been able to attract software talent in large numbers. Bangalore is one of the few cities in India with pubs and an interesting nightlife for Westerners. Other major Indian cities, such as Kolkata, are less attractive. In order to attract foreign interest, Indonesian software company Sigma established BaliCamp, pictured in Chapter 1, in the mountains of tropical island Bali. The estate, located in the middle of rice fields, has a swimming pool and resembles a vacation resort more than a technology campus. But many developing countries have not developed the quality of location and have seen some of their top talent leave and have had difficulty attracting foreign activity. We know of Dutch managers, having to stay in Dhaka, the capital of Bangladesh, who found the city unpleasant for a long stay.

Less important factors

In closing, after presenting eight factors which lead to success in building a software export sector, we present two factors which we have found to be *overvalued* by many observers: intellectual property rights protection and the health of domestic software demand.

Most developing countries have very high piracy rates and poor enforcement of violators. For most offshore clients, piracy is not a key factor. The incredible growth of software exports from high-piracy nations, such as India, China or Vietnam, shows the irrelevance of this factor.[31] Some have argued that piracy spurred development of local software skills which otherwise would have been impossible, given the high prices of Western software products. Piracy may have been an incentive for many companies in developing nations to direct their attention to exports since they could not survive from local sales. In the last decade, most developing nations have moved, somewhat grudgingly, to take enforcement actions against the most egregious violators.

Some have argued that healthy domestic software demand is a key success factor for an export industry. The premise in this argument is that easily accessible customers help to grow and stimulate the software export industry. After all, domestic demand can provide much-needed working capital, and helps companies to develop their processes and technical skills to meet increasingly complicated requirements. However, we do not see domestic demand as a key success factor. For example, India developed a strong software industry relying on foreign demand with weak domestic demand.[32] Most smaller developing countries have insufficient domestic demand to spur the growth of a software export industry.

In fact, a healthy domestic market diverts attention of software companies from exports. Mexico, Brazil, and South Africa have strong software industries with export potential, but strong domestic demand led companies to focus their attention on their domestic markets: these are easier to sell to, they are often less demanding than finicky foreign clients; and finally, they are often more profitable. The top 200 IT companies in Brazil had 2001 revenues of 2.1 billion USD, but less than 10% of that came from exports.[33]

Concluding lessons

- Government's greatest impact in building a software export industry is in two areas in which it has clear strengths over marketplace forces: human capital, through investment in science and technology education, and infrastructure, by creating technology parks.
- National policy-makers are faced with four foci in exporting. Offering commodity programming skills, attracting foreign technology companies to set up software R&D centers locally, exporting software products, or skipping software and focusing on ITES. Most smaller developing countries cannot focus on all these niches simultaneously, but must choose one or two, or risk diluting their national focus.
- There are eight principal factors which explain national software export successes. These can be used as a framework for prescriptive policies and strategies. The eight factors are: government vision and policy, wages and costs, human capital, the industry (concentration, competition, and cooperation), capital, technological infrastructure, linkages, and the quality of life.

11 Marketing of offshore services: the provider perspective

While much of this book is written from the viewpoint of the consumer of offshore work (the end-user), this chapter is written from the provider viewpoint. These are the thousands of small- and medium-sized offshore companies (with a handful of large ones) that are seeking to market their services to clients in roughly 20 wealthy, industrialized nations, along with a smattering of clients in mid-tier nations that are also beginning to shop abroad.

We estimate that there are now some 4000 companies in low-cost countries trying to capture a piece of the rapidly growing offshore market. A handful of these service providers, all Indian, have grown into huge and powerful multinational enterprises. These global Indian firms are competing with firms from the industrialized nations and are successfully attracting large customers. For small- and medium-sized Indian companies however, growth has proven to be more difficult.

The marketing of IT services is even more challenging if the offshore provider is not from India. While IT professionals from most industrialized nations are well aware of the "India brand," providers from other nations are at a disadvantage. Prospective clients may often not consider firms from the emerging Tier-2 nations, such as Mexico, the Czech Republic, or the Philippines. The invisibility is even more problematic for the "infant" Tier-3 nations, such as Colombia, Egypt, Belarus, or Indonesia. However, with the continued demand in the global markets for offshoring, there are many business opportunities for small- and medium-sized providers, even for those from "unknown" nations. All of this is quite new and little has been written about the providers' marketing of offshore services, and thus addressing their needs is the objective of this chapter.

The paradox of marketing a *service*, such as software development, is that the quality of the service can only be judged *after* the service is consumed. This makes the marketing of services both different and more difficult than the marketing of products. Products can be described in technical terms, samples can be sent, and they can be inspected and tested. It is much harder to describe the quality of software services. While it may be easy for a potential customer to assess the quality of a bicycle made in China, it is far harder to verify the abilities of a Chinese company to deliver quality software services. The buyer cannot base the decision on price alone, as different prices may also reflect different kinds of solutions or different levels of quality. Hence building relationships of trust with potential customers becomes even more important, since the trust becomes somewhat of a proxy for quality. But building this trusting relationship requires considerable efforts, especially if the provider is located thousands of kilometers away.

Another difficulty in exporting IT services is the business culture of most of the service provider firms. Many IT service providers are founded by technical professionals, such as engineers or programmers. Their passion is not marketing, but developing software and offering IT services, often with an initial focus on the domestic market. These founders tend to believe that being a company with smart programmers and a competitive price will be sufficient in order to gain entry into a foreign market. Unfortunately, this is not the case and many offshore providers have experienced marketing failures. In fact, all firms can expect to encounter problems before being rewarded with some success.

Perseverance is needed in international marketing. When IndiSpeed (an alias), a medium-sized Indian software company, set up shop in The Netherlands, the Indian business development manager said: "We do not want to do marketing now. First we want to find some customers." He did not realize that in order to find the first client, he must do more than try to sell, and he must invest in marketing. He also did not realize that this would cost a lot of time – time his company could not afford. IndiSpeed had to close down its Dutch sales office after just 1 year – without finding a single client.

Several other offshore providers, including EPAM, one of the largest Russian firms, operated offices in Holland for a short period of time before closing them down. Even Infosys, one of India's most successful enterprises, was hasty: it opened a marketing office in The Netherlands in 1994 and closed it a few years later – thus missing the business opportunities resulting from the Internet-hype, the Y2K problem, and the introduction of the euro currency. Infosys returned again to Holland in 2001, and marketing had to start over from scratch. Even providers from Western countries selling offshore services face difficulties. An example is Xansa, a major British IT services company with a large offshore delivery center in India. In 2003, it decided to pull out of continental Europe after several years of disappointing business results trying to sell its offshore services. Its CEO said: "In continental Europe we do not believe that the marketplace is ready for large-scale outsourcing of IT and business processes that leverage our offshore model."[1]

It is unrealistic to believe that any sales manager, exploring a new market, will generate substantial amounts of business in just 1 or 2 years. Providers need to take a long-term vision and should not give up too easily. They can learn from the long-term view of Indian software giant TCS: it set up operations in the UK way back in 1975. After 30 years of business development activities, it now has 3700 professionals working for British clients.

Providers seeking new international clients face a challenging marketplace: potential clients take time to make a decision, they have many offshore providers to choose from, but occasionally still need to be educated about the offshore option. Some firms were lucky when venturing into foreign countries without any market knowledge, perhaps

- *A truly international outlook for exporting.* They seek out world business opportunities as they strive to enter selected overseas markets. As they build up an understanding of international markets they develop an enthusiasm for exporting.
- *A long-term commitment to exporting.* They realize that it takes time to export IT services successfully and are prepared to wait several years for success. They are ready to allocate financial resources for selected target markets, they try to establish a foothold, and they respond positively to problems. They also expect to make some mistakes before achieving success.
- *Thorough research into new markets and development of export plans.* They allocate a significant part of their marketing budget to market research and marketing planning. They gather as much relevant information as they can first hand. They make a number of planned visits to target markets, either by participating in official trade missions or by attending trade fairs. They try to study market needs, identify potential customers, assess the competition, and determine their possible competitive advantages.
- *An international reputation for quality.* Offering low-cost services is not sufficient. It is vital to understand the needs of the client and adopt a total quality approach. It is important to respond promptly and efficiently to orders and enquiries, and to provide after-sales support if needed.

Exhibit 11.1 Characteristics of companies successful in exporting IT services. Adapted from a report by International Trade Center of UNCTAD[2]

because they had a first-mover advantage. However, the marketplace has matured rapidly, requiring newer providers to carefully prepare their marketing efforts.

What separates the successful exporters of services from those that fail? The four success factors in Exhibit 11.1 capture the foundations for success: an international outlook, long-term commitment, thorough research, and a reputation for quality.

Lessons from marketing strategies of the largest offshore providers

Small- and medium-sized companies can draw several useful lessons from the successful experiences of the largest offshore providers. These largest IT services companies are Indian, such as TCS, Infosys, Wipro, HCL Technologies, Patni and Satyam, and are referred to as the Tier-1 offshore providers. A few of them have already exceeded revenues of 1 billion USD. They offer a very broad range of services and have acquired deep domain knowledge in specific verticals (e.g. banking and finance, insurance, logistics, embedded systems, and telecom). Their growth has been so strong that they now aspire to move into the ranks of the top global IT service providers. Some of these firms employ tens of thousands of staff and attract the best available local talent. The other important offshore countries, China and Russia, have not grown such large and powerful firms. As a matter of fact, several of the largest Indian companies have more employees than the total employment in the Russian IT export sector.

The marketing and growth strategies of these Indian Tier-1 offshore providers have proven themselves. Central to the Indian commercial successes has been proximity to markets. The large firms are represented in the major foreign markets and they hire professional sales people, of which many are now local. An example is Wipro, one of the premier Indian firms. It has eight nearshore development centers in major markets and some 30 global sales offices. In addition, Wipro sometimes partners with local players, such as Accenture, to win business. It services more than 300 clients.

In order to grow their business, initially, the large Indian vendors took advantage of the influential diaspora of Indians working in American and British firms to establish relationships. The providers gradually gained their clients' confidence by having Indian staff working onsite and were increasingly able to shift more work offshore. Again, perseverance was required: it was only since 2002 that offshore work overtook onsite work.[3] Today, these Tier-1 firms are respected partners and have brand recognition among corporate buyers. Their company names appear often in magazines and research reports. There is hardly an IT manager in America that has not heard of TCS or Infosys by now.

Proximity to markets is helping the large Indian providers to anticipate market conditions. Proximity to clients helps them anticipate client requirements and strengthen relationships with clients. To be near their customers, these Indian firms are even moving into new territories. TCS opened its first Eastern European office in 2001 in Budapest, Hungary. In 2004, Satyam established its largest global development center outside India in Melbourne, Australia, and opened its first offices in Canada and Hungary.

These large providers enjoy other advantages. Much of their new work is on a "follow the client" basis, where the client, a multinational, decides at headquarters that their subsidiaries in other countries (e.g. GE Netherlands) should also use the services of the offshore provider. For companies such as Wipro or Infosys, the 10 largest clients are responsible for up to 40% of all their revenues. And since customer satisfaction is high, 70% of their work comes from existing customers. These top tier firms are also able to achieve almost 30% higher billing rates than their Tier-2 Indian competitors.

The Tier-1 Indian firms have yet to win the very large contracts (100–500 million USD), or the mega-contracts that are still going to the likes of IBM or Accenture, but this is probably only a question of time. The Indian firms are also moving into the growth markets of offshore IT-enabled services (ITES), such as call centers or back office work. In addition, they are expanding their offerings up-market, by providing various consulting services. For instance, Satyam started offering services in software process improvement (e.g. Capability Maturity Model (CMM)). Other companies are moving to more complex tasks of design and systems integration, or are offering infrastructure management services.

In order to expand, Indian firms have been buying foreign IT companies. Wipro acquired the energy system division of American Management Systems. For 26 million USD, it got 90 consultants and 50 existing client relationships. In 2004, it bought American consulting company NerveWire for 18 million USD. Offshore service provider

Cognizant bought Amsterdam-based Infopulse in 2003, which had clients in the financial services industry in the Benelux and a staff of 40. The Tier-1 Indian companies are expected to continue buying more foreign firms since many have no shortage of funds. TCS, Wipro, and Infosys have stock market valuations surpassing those of American giants EDS and CSC (reminding us of the inflated valuations of dot.com companies a few years before).

This overview of the successful Tier-1 Indian firms provides several lessons for smaller offshore rivals. There are several factors that contributed to the success from a marketing and growth perspective. Smaller providers can emulate some of these factors as they expand their international marketing. First, in spite of the so-called "death of distance," the Tier-1 firms made sure to establish proximity to markets and clients. They have numerous sales offices and development centers outside of India. They built experience and relationships by working at client sites or close to clients. Their professional staff in Europe and America is increasingly hired locally, and is often lured from non-Indian competitors. Second, the Indian firms made use of their diaspora connections at Western firms. Third, these firms are beginning to differentiate themselves through specialization. Fourth, they are expanding their client base by buying their clients through acquisitions and other business relationships. Fifth, much of their business is through repeat customers, who are also, conveniently, large customers. Finally, Tier-1 firms have succeeded in developing brand recognition: their names are recognized by decision-makers. They are no longer faceless foreign organizations.

Of course, smaller offshore providers cannot replicate all of these business practices. They cannot have a presence in many foreign markets, or set up onshore development sites, or purchase local companies. The next section is specifically devoted to small- and medium-sized companies. What do they need to do if they want to successfully export their services? Finding new foreign clients, as depicted in Figure 11.1, consists of three

Figure 11.1 Major phases when exploring new foreign markets.

major phases: conducting various preparatory first steps, followed by local marketing activities in the selected country, followed by business discussions to close the deal.

The first steps

An aspiring company is ready to export its IT services: where does this offshore provider begin its international marketing efforts?

First, the firm must "know thyself." Several questions should be asked, such as the following: Is the management sufficiently experienced internationally? Can the staff communicate effectively in English? Is the company culture client-responsive? What can the company do to be accessible during the clients' working hours many time zones away? Is the staff technically experienced in critical areas? Are the project managers experienced in delivery from afar, or can they learn? How can the firm convince potential customers it is capable of offering quality IT services?

The answers to these questions should spur some internal actions. To these actions the firm needs to add actions related to professionalism. Clients will not give serious consideration to your firm, an unknown foreign firm, if it does not appear to be professional. Two of the factors that create the image of a "professional" firm are quality standards and, separately, a web site. The widespread adoption of ISO 9001 or CMM-certifications gave Indian firms not only a marketing advantage over their rivals, but also the leverage to raise their rates.

A very different kind of professionalism is the website, which should contain abundant information on the company (profile, history, vision, development centers, infrastructure, and management profile). In addition, it should describe the range of offshore services and the specific domain and technical knowledge. The site should also contain white papers, clients' testimonials, photographs of the facilities, and full contact details. Information should be made available in the local languages of the target markets. The web site must have a "Western" appearance since it is supposed to appeal to potential customers. We have occasionally visited provider websites that are awful to the foreign taste.[4]

In parallel to these early actions, the first steps, which are described in this section, include the creation of a business plan, seeking business intelligence, defining potential markets, and discussing market entry strategies. In addition, working through a SWOT (Strengths, Weaknesses, Opportunities, Threats) analysis is useful in order to assess the strengths and weaknesses of the company in relation to the opportunities and threats in the market.

Creating a realistic business plan

Breaking into new markets requires a strategic approach and detailed preparation. Therefore, it is essential to develop a business plan, which considers major marketing factors and tasks within realistic budget constraints.

Successful offshore providers start on a small scale and build up their business gradually. In their business plans, they first identify the types of offshore services required by the market. Then, after assessing the competition, they define a small number of target markets where they feel that they have business opportunities. The companies then investigate the best ways to enter these target markets. Over time, the business plans are updated and the marketing results are measured and compared to the original plan, drawing lessons in the process.

Given the long lead time before any successful international sales are made, the business plan needs to include a harshly realistic assessment of financial resources. Companies are often too optimistic about their sales predictions and the length of time they will need to generate significant contracts. It may also be acceptable to lose on the first projects if necessary. Companies need to consider alternative sources to fund international marketing other than cash flows, such as bank loans, or export assistance (which is discussed later in this chapter). The financial assessment should consider the firm's service pricing. For example, some providers with sales experience in the US have been disappointed in Europe, because European rates are lower than the prevailing American rates.

Seek business intelligence

As part of your business planning activities, gather business intelligence on competitors from your own country and those from the dozens of other offshore destinations. Besides visiting their web sites, make a trip to offshore seminars or IT trade fairs in one of the target countries. In these venues you will have the chance to assess the strengths and weaknesses of the marketing strategies of other providers from up close. Trade fairs also provide an opportunity to do market research, to meet people, exchange ideas, and open up future channels. Do visit the corporate lectures which take place in parallel to the exhibitions; these are not only informative but also a place for networking.

The following hints can be useful for making the best of business intelligence visits to trade fairs:
- Some trade fairs are huge: plan your tour carefully.
- Talk to the representatives: a trade fair is the easiest way of meeting people in person.
- Study the displays and the presentations to help assess how competitive your firm is.
- Collect literature and other promotional material.
- Take careful notes, otherwise you will soon forget many important observations.

Defining target markets

No provider will succeed by trying to market to all countries at once. The exporting company needs to carefully consider its first target markets.[5] We present two key criteria in determining the best target markets.

The first factor is market size. The USA is the principal market of choice for many large Indian providers due to the enormous size of market demand. But not all Indian firms looked at the USA: since its inception in 1982, the focus of Indian-based Mastek has been the UK and it is one of the few Indian-based vendors whose board members include a British executive. When targeting the two largest markets for offshoring, the USA and the UK, some sales managers can generate 3–4 million USD of business per year. However, these large markets are also crowded with many competitors. In the UK alone, hundreds of offshore services providers are trying to find customers.

Although often overlooked by offshore providers, small countries can be attractive because competition is less fierce. An example is The Netherlands, which is a medium-sized IT market. Dutch companies have been using the services of offshore providers for more than 20 years and are relatively open to this approach. The Dutch, more than the French or Germans, have an international outlook and are often willing to cooperate with foreign companies. More than 250 Dutch firms have had their software developed in India alone. Software contracts have gone to at least 35 other offshore destinations. Research in 2004 indicated that around 5000 offshore staff members were involved in projects for Dutch companies. This number is estimated to grow to 50,000 in 10 years; a huge volume for such a small country.[6]

The marketing factor that has proven most successful for many offshore firms is the use of *linkages*. These linkages result from linguistic, historic, ethnic, geographic, or emigration-related reasons. Based on geographic proximity (nearshoring), the US is a logical market for Mexican firms; Western Europe is a logical target for Central and Eastern European companies. Using linguistic links, some Indian firms are using French-speaking staff in Mauritius to target France. Spanish-speaking companies in Latin America are focusing on Spain. Linkages are also discussed in Chapter 10.

Another linkage is the diaspora. Many nations have sizable diasporas in key target markets. These diasporas can play critical bridging roles in the early stages of business development. A study on small- and medium-sized American companies that offshored to countries other than India revealed that in nearly every case, it was a Pakistani, Indonesian, Vietnamese, or other foreign national working in the client company who pointed to the offshore partner.[7] Similarly, in nearly all cases of successful Iranian software exports, there was some involvement of Iranian expatriates.[8] Not all offshore providers are taking advantage of such linkages. For example in The Netherlands, which has a large Turkish community, one finds a number of Turkish-born people in the IT sector, yet no Turkish companies have yet mobilized them to foster business links.

A local base is desirable

It is extremely difficult to find clients if a company is located on another continent, thousands of kilometers away. Even in Europe, establishing a foothold in one country, such as the UK, will not give easy access to other European countries. Most clients feel more comfortable communicating with their offshore provider with a simple local telephone call rather than having to arrange communications across time zones. A local base is therefore a marketing base, a base for relationship building, as well as a base for liaison activities once the work is underway.

The most effective way for an offshore provider to achieve proximity to markets is to go all the way and open its own sales offices abroad. Only with a local presence, ideally with local staff, can marketing and sales be most effective. A local base makes personal contacts and personal visits easy, and enhances the understanding of the market. It is also an expensive way of doing business, since office space must be rented and a business development manager needs to be paid at Western levels, considerably above any wage levels of the rest of the offshore organization.

If it is not financially possible to establish a sales office, the offshore provider needs to consider other means of local representation. For example, working with a local representative or agent is a less-expensive alternative. A capable local representative offers market knowledge and has a network of contacts. However, the representative will usually request a fixed fee, in addition to commission, to conduct appropriate marketing and sales activities; while the offshore provider will prefer to work on a "commission only" basis, where the agent only receives money on signed contracts. Finding foreigners willing to work on such a "no cure no pay" basis is uncommon, although one of your country's expatriates may agree to such an agreement.

Some nations have set up a joint office representing a group of offshore companies, a consortium of companies, or the entire association company membership. This approach significantly reduces the market entry costs for offshore firms. An example is the Bangladesh ICT Business Centre (BIBC), which was set up in 2003 in Silicon Valley. The goal of this shared sales office was to promote and speed up direct contacts between 30 Bangladeshi providers and potential American clients. Initiatives like this are promising but have had mixed results: the Brazilian government created Softex (the Brazilian Society for the Promotion and Export of Software) in 1992, followed by the opening of foreign offices in the mid 1990s. However, due to disappointing results, these were closed down a few years later.[9] Such collaborative efforts are unlikely to work unless large-scale and aggressive marketing takes place.

Another approach for offshore providers is to seek a business relationship with a local IT services company in one of the target nations. Many of these domestic firms are eager to offer offshore services in order to stay competitive, but do not want to set up their own captive unit abroad. The local company will handle marketing activities, reducing the risks substantially. For example, Indian-based HCL Technologies accessed the Dutch market by partnering with Roccade, a Dutch IT services company, in the 1990s. At that time, it did not have its own Dutch marketing office.

The offshore provider should also try to establish contacts with offshore intermediaries. These are consultancies and research firms that provide expertise and services to firms seeking offshore work. Some of these intermediaries, especially in the US, have amassed enormous influence. It is useful to make your firm known to these intermediaries, because clients look at them to select the offshore provider.

Finally, another form of intermediary is the *online marketplace* that provides programming matchmaking.[10] Hundreds of offshore firms, most of them small, use these marketplaces as their primary marketing channel. Many of the projects presented for bids are quite small. Offshore providers use these small projects to "buy" their first clients and to gain experience. But more important, they are buying favorable "reputation marks" on these sites, which allow them to slowly build relationships and progress to larger and larger projects. These online intermediaries live up to the promise of giving "global reach" to every small business.

Conducting a SWOT analysis

When considering a new market, it is useful to conduct a competitive analysis of your company. A *SWOT analysis* covers the following:

- *Strengths*. Where does your firm excel? These are the factors that will catch the attention of a potential client and persuade it to buy your services. For example, one small Russian firm was considering entering the US market. "What are your strengths?" we asked. At first, the co-founder gave the usual response: that the firm's engineers have many skills and build web pages. "But how are you different than all the other firms?" we asked. "We have strong expertise in computer modeling," was the reply. Here was their strength.
- *Weaknesses*. Where is your firm lacking relative to your competition? These weaknesses need to be addressed before your firm markets internationally. For example, a common weakness for many offshore providers is small size and limited financial resources. If you cannot afford to set up your own sales and marketing offices, you could consider using expatriates or agents.
- *Opportunities*. These are external factors of which you could take advantage. For example, there may be some new market demand for software services, such as the Y2K remediation need in the late 1990s.
- *Threats*. These are external factors that could harm your firm if not counteracted. Most threats are from other competitors – other offshore providers.

OriginalJava (an alias) is an actual Indonesian IT services firm that considered entry into The Netherlands (Exhibit 11.2). The company has 200 employees and is located in the capital city, Jakarta. A SWOT analysis, which is typically depicted as a 2×2 matrix, is shown in Figure 11.2. OriginalJava listed six strengths in the SWOT matrix, but its management felt that the two most important strengths were its very low costs and its considerable expertise in Java (the computer language, not the Indonesian island).

The managers felt that the many opportunities in the Dutch market were real, especially those that capitalize on the Dutch–Indonesian links. However, they recognized

Strengths	Weaknesses
• Specialized in Java skills. • Very low rates. • Robust scale of local business (both Indonesian clients and multinationals). • Some nearshore clients in Australia, Malaysia, and Singapore. • Stable company. • Good English skills.	• Completely unknown in The Netherlands (or elsewhere in Europe). • No customers (or references) in The Netherlands. • No CMM certifications. • Few international business alliances. • Marketing and sales staff is predominantly Indonesian. • Country branding: Indonesia is unknown in the field of IT and its political unrest creates a negative image.
Opportunities	Threats
• Historical links between The Netherlands and Indonesia (a former colony). Indonesian cuisine is the most popular foreign food in Holland. • There are tens of thousands of Indonesian expatriates in The Netherlands. • Dutch managers are under cost pressures. • Dutch companies need flexible staffing solutions. • Dutch providers are interested in alliances and other partnerships. • Dutch companies are seeking offshore ITES and some Indonesians speak the Dutch language.	• Indian competitors are already active in The Netherlands. Some have their own sales offices; others operate through agents. • Nearshore providers (from Central and Eastern Europe) are also present in the local market. • Several Dutch services companies are also offering offshore services.

Exhibit 11.2 SWOT analysis of an Indonesian offshore provider considering entry to the Netherlands market.

that the firm had many weaknesses, particularly the lack of business contacts in The Netherlands. Although the firm was stable and had positive cash flow, management was reluctant to invest in a full-time local marketing office, and considered alternatives, such as using the Indonesian diaspora or hiring local marketing consultants as agents. After much deliberation, the company decided to test the waters by appointing a member of the Indonesian diaspora as a sales agent. His responsibility was to conduct the marketing and sales activities for a trial period of 18 months.

Local marketing activities

Having a local base is often the first step to successful business development, but even then, new clients will not come automatically to your door. To win new customers,

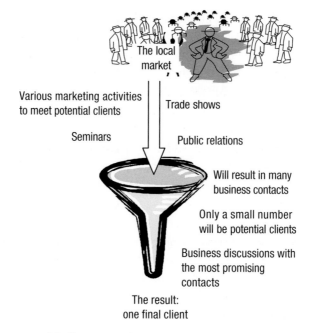

The local market

Various marketing activities to meet potential clients

Trade shows

Seminars

Public relations

Will result in many business contacts

Only a small number will be potential clients

Business discussions with the most promising contacts

The result: one final client

Figure 11.2 The process of finding a new client.

conducting effective marketing activities on the local level is required. In this section we introduce several important elements of marketing at the local level derived from our experience.

As depicted in Figure 11.2, finding customers is basically a "numbers game": the larger the number of relevant business contacts, the easier it will be to find the first client. In most cases, of the many new business contacts you are able to make, only a small number will be relevant. These are the potential clients. And business discussions with many of these promising contacts are needed before the first client can be found. Making the name of your company known (establishing name recognition) through marketing efforts accelerates this process, but identifying and meeting potential clients requires a lot of time. It can take a long period – sometimes years – before a substantial contract can be signed. Sales cycles are especially long in the banking and financial sector.

As a provider you are anxious to generate client leads. There are many ways to meet potential clients, such as the traditional means of gaining access through references, networking, direct marketing, and cold calling. We focus on efforts that require special attention by offshore providers: large-scale marketing efforts, specifically trade fairs and seminars; and separately, public relations.[11]

Trade fairs

We know some companies that have found most of their clients through participation at trade fairs, for example at the CeBIT in Hannover, the largest IT fair in the world.

At CeBIT 2004, a large number of offshore providers participated. They came from Central and Eastern Europe, but also from Bangladesh, Brazil, Egypt, India, Indonesia, Iran, Jordan, Lebanon, Malaysia, Mexico, Morocco, China, Sri Lanka, Thailand, and Tunisia. Companies also participate at fairs such as OutsourceWorld in London and New York, and Gitex in Dubai. Providers that are specialized should consider taking part at a specialized (vertical market) trade fair, say, for embedded software, banking and financials, or health care systems. Providers that need to minimize costs participate as part of a country pavilion, which is cheaper than having one's own stand.

There are only a few seconds to attract the visitors' attention as they pass by, so the stand should be professional and attractive. Graphics should be simple and easy to see from far away; the message conveyed should solve a problem. The provider needs to qualify the visitor politely before starting discussions. All this requires special skills that can be learned at specialized seminars on trade fair participation.

Follow-up on all leads after the trade fair is over (remember that marketing is a numbers game). You can follow-up by mail, telephone, or personal appointment. Sometimes, though, well-intentioned follow-up can go terribly wrong. As a friendly gesture, an Indian provider added a small piece of sweet-smelling incense in the envelope along with its follow-up letter. This turned into powder by the time it reached the foreign destinations. The mailing was sent out just after the 2001 attack on America and the foreign letters containing an unknown white powder caused panic among some of the recipients.

While trade fairs have been successful for some offshore providers, others have not been so lucky. The fairs are time consuming, extremely tiring, and quite costly. Contracts are rarely signed at trade fairs. In many cases, the results will not come after the first participation. Visitors may see your stand, but it will take two, three, or more events to generate clients. An Offshore Outsourcing Exhibition in Amsterdam presented more than 60 firms from Russia, Belarus, Slovakia, India, Pakistan, Jordan, and Vietnam. The organizers failed in attracting visitors and some participating companies left the two-day event without collecting even one business card. A lesson from this is that before you decide to participate in a trade fair, always check on the number of visitors.

Seminars

Seminars are more focused, smaller, and usually less costly alternatives to trade fair participation. Setting up a stand at a seminar may not cost much, while becoming a sponsor will give your firm visibility by having your company name printed on all marketing material. Although it will easily cost several thousand dollars, sponsorship will make your firm known, even among people that will not visit the event. You can also organize brief seminars yourself, such as a business breakfast, an afternoon seminar or a dinner. These should be done tastefully and be informative about the offshoring process. Bring in some outsiders to speak who will give color to the seminar, such as one of your customers, a local professor, or a local consultant. These are perfect occasions to meet potential customers in a more informal setting, and they can also be used to invite journalists.

In all cases, the importance of promotional efforts should not be underestimated. Promotion is not always fully understood by the providers. A number of offshore seminars have been hosted in The Netherlands, with delegations from countries such as Bangladesh, the Philippines, or Hungary. However, marketing efforts were minimal and the foreign delegations were sometimes larger in size than the audience. Professional Dutch organizers usually send out more than 10,000 notices to announce an offshore seminar; this large number might result in 50 participants. In order to share costs, you could collaborate with others, such as local firms or even competitors.

Public relations

In the offshore business, advertising is expensive and is unlikely to generate returns. Free publicity is a more effective way of drawing attention to your firm. This means that contacts must be made with journalists in the target market. Come up with a newsworthy subject so the journalist can easily write about your company. You can issue regular press releases to trade publications and local papers, but these too require a newsworthy subject, otherwise they will be ignored. The press release may include an invitation for one or two journalists to visit your country and your facilities. If the local IT association organizes a press tour, then costs can be shared. The British newspaper Financial Times regularly carries an issue focusing on the Indian software sector. The articles are written with support from the Indian software industry association and from individual Indian companies.

There are also foreign journalists stationed in many offshore destinations. Although they do not focus on IT, you could invite them for a visit to your offices: the more attention in the press the better. Appearing on television can be very effective: a Dutch services provider with a Romanian offshore facility gained several clients after it appeared for 8 minutes on a TV business channel.

Of course, the large offshore providers already benefit from professional public relations firms, which manage their "free publicity." They are also expanding into more esoteric branding: Cognizant sponsors players on a British cricket team. TCS and Mastek created links with local universities to foster research on offshore outsourcing.

Dealing with prospective clients in business discussions

The various marketing activities described above result in business leads and business discussions. Business discussions are sometimes straightforward and a contract can be signed in a short span of time. With larger companies, or larger projects, this process is more complicated. In this section we introduce different client types, client cultures, and issues of ethics and trust.

But, first comes *price*, namely what rate (the base hourly rate) to charge. After all, the business discussions are about offshoring – and offshoring is driven largely by price. Some clients may say: "your rate is too high." Perhaps the rate is actually too high, as some providers have become used to American rates, and they overlook the lower price levels in Europe. Alternatively, many providers try to penetrate foreign markets by asking low rates. It indeed makes business sense to "buy" the first client (or even the second client) by offering a low rate. But this may send a bad signal. The prospective client may think "the rate is too low" and therefore the provider's quality, professionalism, or productivity is poor as well.

Knowing your client types

There are various types of client organizations which fall along the continuum of the Offshore Stage Model (introduced in Chapter 1): from those organizations that use little to no offshoring to those that are already committed to offshoring.

Regardless of the stage, your task is to identify and interact with the individuals that are the "offshore champions." These are the influential individuals inside the organization who are already committed to offshoring as a solution to organizational needs. There is no need to "sell them" on the offshoring approach. They are already sold. These individuals are to be found in many organizations – even the ones that have not begun offshoring.

The most difficult discussions are with those managers who are not committed to offshoring. These prospective clients are usually found within companies in Stage 1 of the Offshore Stage Model, labeled as the "Offshore Bystanders." These firms, which are the majority of the companies in the industrialized countries, do not have any offshore projects yet, and may be reluctant to use offshore providers. Managers in these companies are aware of offshoring, but tend to see mainly the negatives. The common objections in this group include:

- "The communication will be difficult."
- "Your project management skills are weak."
- "I'm worried about my Intellectual Property."

Be ready for these objections. Be ready in more than just words by having established the management and legal processes in place to address them. Some prospective clients may not voice all of their objections openly: they have negative views regarding certain offshore nations, or they may be insecure about their own lack of experience with international projects, or they may be sensitive to internal resistance and backlash.

Make them feel comfortable with offshoring by using a goal of starting with one low-risk pilot project. Take your time to supply managers of these companies with sufficient information, such as case studies or white papers. It helps if some documentation is available in the local language. You might invite them for a visit to your facilities, so they can have discussions with your management and staff, and have a better understanding of your capabilities. Add some tourist excursions and you will make an impression they will never forget.

Other companies have transitioned to the second of the offshore stages: the "Experimental Stage." These organizations, the Experimenters, are already offshoring some of their work and there is little doubt that they have some offshore champions. However, they have not structured their provider relationships. The Experimenter's approach to provider selection may be *ad hoc* and it may have several offshore providers that churn, or are otherwise not well-established inside the firm. Even if the Experimenter has designated "preferred suppliers," it is often open to new services and locations, perhaps as a diversification strategy. All of these openings present an opportunity.

The Experimenters require different marketing efforts. The challenge is to increase their use of offshore resources. To achieve this, the offshore provider must help these clients in achieving on-time and on-budget projects, and share best practices with them. Problems in dealing with different cultures can occur and vendors should take a lead in proposing inter-cultural training: they should not wait for the client to address this issue. Vendors should publish newsletters and organize "customer days", seminars or advanced workshops on various topics, such as best practices, international project management, and on other off-shore opportunities, such as ITES. User groups should be invited to the offshore facilities.

Remember that your client base represents your best clients and repeat business is easier than chasing new clients. A golden rule for sales and marketing functions seeking to find foreign clients is that long-term relationships are crucial. Stable growth comes from repeat business; repeat business comes from relationships. A survey by Forrester noticed that "One of the most important areas of differentiation lies in the vendor's engagement and relationship management philosophy, and overall relationship management skills."[12] Some providers have relationship managers, who act as an intermediary between the client and the offshore development teams. They help their clients in their efforts to get a larger percentage of work offshore.

Knowing your clients' cultures

We have already devoted an entire chapter to the impact of cultural differences on how we communicate (Chapter 9). The emphasis in that chapter was on communication during the project life. Of course, these cross cultural communication differences also affect the way business discussions are conducted between a sales representative from an off-shore provider and the prospective client. One of the marks of maturity of the Indian Tier-1 firms is that they are increasingly hiring locals for client relations: Germans in Germany, Americans in the USA. Such marketing choices reduce cultural differences.

For those offshore providers which choose to send one of their citizens to conduct business development, consider training that deals with the specific European (business) cultures. We noticed that providers' marketing and sales managers, with business experience in the USA, expect that marketing in Europe will be the same. This is not the case, and European markets are fragmented across borders. From a cultural point of view, the European markets can roughly be divided between a "Northern" and a "Southern" part, divided by a border which runs across Belgium. And even inside these two distinct parts, cultural differences exist.

For instance in The Netherlands, hierarchy is not easily visible and people appear to be easy-going during business meetings, which always begin with informal talk. Managers from the offshore vendor are misled by this casual attitude and expect a business agreement after having had a pleasant meeting. However, this is rarely the case. Also typical for the Dutch business culture is the long time to arrive at a decision, which is only made if all relevant people and departments agree. In Germany, which also belongs culturally to the "Northern" part of Europe, organizations are more hierarchical, and the people appear more formal and distant than the Dutch. In meetings, they will get straight down to business. Issues of quality are extremely important and sales staff should be well prepared on this subject. In the "Southern" countries of Europe, such as Belgium, France, Italy or Spain, business is hardly possible without good personal contacts. The business culture is less formal. Here, too, business decisions can take a long time, although unlike the Dutch who need to build internal consensus, the time is required in order to build trust with the offshore provider.

Ethics and trust

Like many sales people before them, over-eager offshore representatives have been known to claim capabilities and expertise which they do not actually have. Boasting "we can do everything" is risky. If the contract is signed, the chances of failure are high. This will hurt your firm's future options. There is a saying: "trust comes on foot, and leaves by horse", meaning that the reputation of your company can be damaged quickly. Intellect, the British Information Technology Telecommunications and Electronics Association, has a code of conduct for its members that are offering offshore services.[13] The code gives guidelines on presenting true information on capabilities and staff. It also deals with issues such as respecting local immigration rules, protection of intellectual property rights (IPR), and confidentiality of clients.

On the other hand, providers also need to be cautious about their prospective clients. The provider should be slightly suspicious of requests for proposals (RFPs), especially if they require great effort. Sometimes, the client has already found an offshore partner, and only requested the participation in the proposal because it needs data for negotiations with the preferred firm. An Indian sales manager recalls:

> *"Over a longer period of time, we had regular meetings with a potential Dutch client. Since it was a large company, we had no hesitation in traveling from our office in the UK to The Netherlands. Afterwards, we found out that they had no interest to work with us, since they already had identified another Indian vendor. They only had discussions with us to learn from our experiences and to check on prices. They used this information to negotiate with the other Indian company."*

The customer is certainly allowed to conduct due diligence on your firm, but you are allowed to do this as well: Is the client competent enough to provide you with good and stable specifications? Will the client keep changing the requirements? Is it possible that the client will blame you when the project fails as a result? A client may cause damage even if your firm has acted reasonably and professionally.

Country branding – marketing your country

Tell me what country you are from and I will tell you how easy it will be for you to do business here.

If your company is not from India – and particularly if it is from a small nation – then your first concern is often to "market your country." So, let's briefly examine the case of India and draw some lessons from its success at country branding in IT.

In 1995, most business people knew nothing of India's IT capabilities. The country was known as a large, poor, and hot country, with images of elephants and snake charmers giving it an exotic aura. This view has changed dramatically and today, Western IT professionals know India as a potential high-tech destination. Offshore outsourcing and India have become almost synonymous. This new image has had an important positive impact on the business development activities of Indian providers. For example, most Dutch companies, when deliberating on selecting an offshore partner, limit their selection process to Indian firms.

On every level, from government, to state, through industry coalitions, to individual firms, India has carried out a successful national marketing effort to build its image and its brand as a leader in information technology. The enormously successful major Indian providers have all helped build the image through substantial financial investment. The glittering campus of Infosys not only helped Infosys but also helped the India brand. The many activities of the national government and its ministries have aided the branding. Even Indian states have their own promotion policies: Andhra Pradesh (home of Hyderabad) and Karnataka (home of Bangalore) can be found at trade exhibitions abroad.

Of particular note within India's strong branding efforts is its aggressive national software industry association, National Association of Software and Service Companies (NASSCOM). It is perhaps the most influential technology association in the world. Founded in 1988 with only 38 members, it has grown into a large organization with a variety of activities. The combined revenues of its 850 member companies constitute almost 95% of the total revenue of the Indian software industry.

From its early days, one of its aims has always been to strengthen the branding of India as a premier global sourcing destination. It welcomes foreign delegations and sends out Indian IT missions abroad on a regular basis. Its website is a rich and reliable source of information. Lastly, the association has been successful at facilitating favorable studies about the Indian industry. An example is the 500 page World Bank funded study of 1992, which compared India with seven other countries.[14] Indian companies and organizations communicated the positive findings of the report extensively in subsequent years. In 1999, NASSCOM began working together with McKinsey, a well-known strategic consulting firm. This relationship has proven especially fruitful as McKinsey regularly publishes favorable research reports on India's IT industry.

NASSCOM's leadership has been instrumental in its success. Dewang Mehta headed the association from 1991–2001 and became a legend. It was his belief that software and IT services export was to India what oil was to the Middle East. He was creative in building NASSCOM into a professional organization and was personally very active in promoting the Indian IT sector. He gave countless interviews to local and foreign magazines and newspapers, and participated in a large number of seminars and trade fairs abroad. He died in 2001 during a business mission to Australia. The Indian Ministry of Information Technology instituted an award for IT innovation to honor his memory.

India's country branding stands in contrast to the image of many other competing nations. The image of most developing countries is often negative (e.g. poor, uneducated, with a bad infrastructure). For example, Bangladesh appears at the top of corruption surveys.[15] Although corruption has little impact on its offshore projects (as foreign users will admit), it is obviously negative from a marketing point of view. Some Central and Eastern European nations suffer from a poor image as well. Other nations are better known because they are holiday destinations (e.g. Malta, Jamaica, Cuba, Egypt, and Indonesia), but their IT capabilities are invisible.

Since the end of the 1990s some offshore countries began promoting their countries as IT destinations. Nepal is famous for its tall mountains, but hardly anybody is aware of its IT offerings. Designco and World Distribution Nepal, offshore providers from Kathmandu, are now using the slogan "software from the top of the world".

Another example of a country branding strategy is the Baltic nation of Lithuania. It aspires to become the "Sunrise Valley of Europe" by 2015 in information and communication technologies, laser technology and biotechnology.[16] Like most Eastern European nations it tries to emphasize its cultural closeness to its potential European customers in order to distinguish its industry from the "alien" cultures of India and China. In order to achieve its goals, Lithuania has to project its strengths and capabilities.

In creating such a brand, Lithuania operationalized its branding strategy in the following ways:

- Use aggressive communication in order to be heard over all the competing and non-competing messages being sent out by other countries.
- Use a broad-based campaign, where CEOs of local companies would focus on addressing events and using media. Strong public relation firms should be used to promote success stories of Lithuanian companies.
- The local IT association, Infobalt, and the Lithuanian Development Agency should work together to focus on the mission "Lithuania as the Sunrise Valley of Europe."

- Demonstrate successful implementation of niche areas both domestically and abroad. This would lead to assurance that Lithuania is able to deliver quality projects and products.

This example of Lithuania and its Infobalt association demonstrates the subtle role that the national association plays in building the country branding to compete with the India brand. Industry associations have formed in most offshore countries, such as Camtic (Costa Rica), Fedesoft (Colombia), Pasha (Pakistan), Basis (Bangladesh), CAN (Nepal), Intaj (Jordan), PIIT (Poland), Asocpor (Czech Republic), BAIT (Bulgaria), Litta (Latvia), Russoft (Russia) and IT Ukraine. Some of these are also members of international associations.[17] But, unlike India's association, few of these industry associations are visible outside their home nations. Some associations suffer from internal weaknesses, and in some countries, such as Romania and Russia, there has been rivalry between or within industry associations.

An association also has responsibility to promote high standards of professionalism and ethics among its members. A large firm from Ukraine attended an offshore conference in The Netherlands, but it never paid entrance fees. Its representatives also used the copy machine extensively to make additional company brochures, without paying. Bad news travels fast, and such a negative incident damages the image of an entire country's industry.

Associations can offer their member firms tangible benefits. They can organize member participation in conferences and exhibitions in foreign markets. They can organize training (e.g. on export marketing, software quality, or project management). They can be a useful source of business intelligence. Some industry associations host exhibitions on a regular basis, which present the country's IT capabilities in a good light. Examples of annual events are Infobalt (Vilnius, Lithuania), Software Outsourcing Summit (St. Petersburg, Russia), ITCN Asia (Karachi, Pakistan), and the Jordan ICT Forum (Amman, Jordan). While these events attract few buyers, they attract influential foreigners, such as consultants, who help spread the word.

External assistance with market entry

A surprisingly large number of outsiders are eager to help offshore providers from developing countries (and transition economies of the former Soviet Union) to make the important and expensive first steps in exporting their services. Support can be sought from governments of wealthy nations and international agencies. Governments from several offshore countries are operating trade promotion organizations in foreign markets. They sometimes fund trade missions, seminars, mailings, and support national booths at trade fairs.

Several Western European governments operate export promotion projects. An example is the German-sponsored Indo-German Export Promotion (IGEP) Project.

This is a joint trade promotion program of the Indian Ministry of Commerce and the German Ministry of Economic Cooperation and Development. It provides support to Indian offshore providers in establishing contacts with German and European enterprises, and assists them to organize trade exhibitions and seminars. The REACH program of Jordan receives support from Switzerland. The Dutch have CBI, the Centre for the Promotion of Imports from Developing Countries, which has assisted offshore firms from India, Jamaica, Nepal, and Bangladesh to take part in foreign IT exhibitions. It also offers training, disseminates information, and produces a European Market Survey.

The European Commission has assisted a number of firms from developing countries in gaining access to European markets. One such program was 3SE (Software Services Support and Education Centre), which was a joint initiative of the European Commission and the Government of India, with offices in Bangalore (India) and Brussels (Belgium). Its aim was to promote cooperation between the European Union and India in the field of IT. From 1999 to 2003, 3SE offered matchmaking services to help European companies find the right partners in India for software services and products. Seminars and matchmaking events were organized, both in India and Europe. It also helped to distribute European software products in India.

European-funded promotional projects have taken place in many countries. In India, the industry association, NASSCOM, initiated the NASSCOM's India–Europe Software Alliance (NIESA) project in 1999 to increase strategic alliances, joint ventures, and partnerships between companies in India and Europe. This was followed up in 2000, with the NASSCOM's India Japan Software Alliance (NINJAS). Another example is the European IT Service Center (EITSC) established to increase the software exports from the Philippines. Yet another program, called New Adonis, was created to support the software industries of Armenia, Russia, and Ukraine by searching for clients in Europe.

American governmental agencies have also been assisting offshore providers. For example, the US Chamber of Commerce has been helping software and IT firms from Thailand seek American clients. The various bilateral partnerships of the US Chamber, such as the US ASEAN Association or the US Indonesian Trade Association, have provided marketing assistance to visitors from the countries they support. The US State Department has sponsored a CIS forum of software companies including Russia, Armenia, Georgia, and others.

USAID is involved in a variety of activities with the IT sectors of many countries, in order to promote foreign investments and to attract new clients. It hosts delegations, it funds firms participating in foreign IT exhibitions and helps to attract venture capital. It gives assistance in developing business plans, and offers training to strengthen marketing skills (e.g. with developing marketing plans and preparing elevator pitches).[18] USAID also supports various local IT associations, including those of Palestine and Jordan. It supported the efforts of several Arab firms to develop software for the lucrative Arabian Gulf markets, by providing seed money to translate standard software packages into Arabic.

Finally, various international organizations have helped companies in developing nations: the United Nations Development Programme (UNDP), United Nations Industrial Development Organization (UNIDO), the International Trade Centre UNCTAD/WTO (ITC), and the infoDev program of the World Bank. Regional development banks such as the Inter-American Development Bank (IDB), Asian Development Bank (ADB) and the European Bank for Reconstruction and Development (EBRD) have all provided technical assistance to firms in their regions.

Concluding lessons

- Providers should not give up too easily. The international marketing of offshore services is complex and time consuming. For most markets, expect 2–3 years before significant business will be generated.
- Proximity to markets is key. It is difficult to find clients if you are located thousands of kilometers away. It is important to invest in a sales office in your target country or to use a local representative. In addition, seek a business relationship with a local IT services company, and establish contacts with local consultancies and research firms.
- The power of the diaspora is often underestimated. Capitalize on your network of expatriates working in potential target firms (such as, for example, a Filipino provider targeting a Filipino tech manager at a Canadian health care organization).
- Finding customers is a "numbers game": the larger your number of relevant business contacts is, the easier it will be to find the first client. Many different types of marketing activities are needed to generate name recognition and client interactions: references, networking, direct marketing, trade fairs, seminars, and public relations.
- For non-Indian providers country branding is critical. Individual firms, government ministries, and IT associations need to actively work together in conducting country marketing efforts.
- Seek support from one of the many programs funded by governments from wealthy nations or from international agencies.

12 Offshore politics

Backlash:

A sudden violent backward movement or reaction;
A strong adverse reaction (as to a recent political or social development).

Offshoring has become one of the most important social issues of the early 2000s. While there were muted political discussions in the 1990s, the catalyst to the backlash was the Forrester report of November 2002 in which this American research organization forecast that the US would lose 3.3 million IT and office jobs to offshore destinations by 2015.[1]

For the first time, jobs that were migrating offshore were not in factories, or assembly lines, farming, or mining, but those of white collar, highly-educated professionals. This was new and quite unsettling. We have not seen this before.

Across the US and Western Europe programmers were losing jobs. Labor unions were responding to outsourcing with strikes or strike threats, such as those of the IT departments of the Bank of Ireland and of Swansea County Council in Wales (although in both cases no offshoring was involved). UK-based financial giant HSBC did face a strike due to offshoring IT-enabled services (ITES) positions in customer services. The American edition of Computerworld published a photo of a protesting programmer, with a sign reading "Will Code for Food." Forrester followed-up its famous American report with a forecast that the UK would lose 750,000 jobs to offshoring by 2015, of which 150,000 would be IT jobs. In Australia offshoring was forecast to cost 40,000 computer jobs by 2015.[2] The issue of offshoring became a crisis. In America, TV commentator Lou Dobbs, of CNN, relentlessly attacked companies that were offshoring. Outsourcing had become a dirty word.

The immediate policy issue: job loss and wage decline

The estimates of job losses that began emerging were troubling. Prominent in highlighting the crisis was Ron Hira, of the IEEE-USA, an association of engineers, who pointed out that in the 30 years that the US government has been collecting such data, the years 2001–2003 were the first in which electrical and computer engineers had higher unemployment rates than the rates for all workers.[3] Another study found that within 1 year of the peak of the tech boom, US employment in IT industries declined 20%.[4] Yet another study found that unemployment rates for US programmers reached 6.4% in 2003.[5] IBM announced in late 2003 that it would replace 3000 US employees due to offshoring, although it later reduced this number due to bad publicity. Britain-based telecommunications services provider Colt Telecom announced that it would transfer some 300 jobs in Germany to India by the end of 2005. The company announced that it would offer 20 of its top managers the opportunity to shift to New Delhi (although at Indian wage levels).

Globally, Deloitte Consulting predicted a loss of 2 million jobs offshore over the next 5 years in the field of financial services.[6] Even Finland's miracle company, Nokia, laid off workers in 2003, as the company offshored to Estonia, India, and China.

> In late 2003, Business Week created a stir when it told the story of a small Boston-based firm that was deliberating whether to hire four offshore programmers for a fast project. The firm's director was torn between business reasons and "moral" issues. Finally, he thought of a compromise. He would place an ad in the leading local newspaper, the Boston Globe, advertising for a US-based programmer, but at roughly Indian charge rates plus a small premium (45,000 USD). He was flooded with 90 resumes, many from excellent programmers.[7]

As the Boston story illustrates, for the software professionals who continued to be employed, software wages were standing still, or in some cases, decreasing. One source reported that computer wages declined 20% in the 2002–2004 period.[8] There was concern that offshoring would create long-term IT wage reductions in the USA and Western Europe.

Reactions to the backlash

In the months and years following the 2002 Forrester forecast, two camps emerged (in America and elsewhere): the first camp saw offshoring as a crisis that requires immediate action to repair. The other camp, positing free trade arguments, argued that offshoring is inevitable and that it will benefit all industrialized economies through greater efficiencies. The two camps have been labeled "The sky is falling" and "Don't worry, be happy."

Reaction 1: The sky is falling/protectionism is needed

The reaction in US politics to dire forecasts was swift. By 2004 more than 20 federal law proposals ("bills") to restrict offshoring were proposed in the US Congress. One of these proposals, made by Senator and Presidential candidate John Kerry, would require disclosure of location of a call center at the beginning of a call ("Hello my name is Raj and I am calling from Kolkata, India"). One of the proposals, passed by the Senate, although unlikely to become a law, stipulated that government contracts will not be granted to a company that replaces US workers with offshore workers, if such work was previously done by US workers (even if passed, this law is unlikely to stand up to the US' World Trade Organization agreements). Other proposals, such as that of Senator Hillary Clinton, set out to limit offshoring by limiting personal data exporting, similar in its impact to the European Union Privacy Directive that took effect in 1998.

Of the 50 states of USA, 36 states have discussed some kind of legal restriction on offshoring. All the Democratic Presidential candidates of the 2004 elections developed political platforms and proposals on reducing "outsourcing" (the code word for offshoring).

In Europe, concerns soon followed those voiced in America. For example by late 2004 the French government set aside a 1 billion euro fund to entice French companies to continue operations in France if they guarantee not to move jobs offshore.

Reaction 2: Don't worry be happy/free trade is good

The opposing camp did not see offshoring as a problem that needed to be fixed. At the time the political debate was ignited, President Bush's administration in the US espoused this view and chose to leave trade open and not to implement protectionist steps. Across the Pacific, the Australian Trade Minister went even further in 2004, voicing support for offshoring large portions of Federal IT work to reduce costs.

This camp argued that the new fierce competition in services is simply a natural outcome of global trade liberalization. For Americans free trade has been positive, raising standards of living, and it has been positive for growth in the long run. Americans are beneficiaries of the free flow of capital and investment with millions of Americans employed by foreign firms.

The number of jobs lost to offshoring is very small compared to the "creative destruction" – the annual churn of America's flexible labor markets which turn over two million jobs per month.[9] The US Department of Labor, the most respected American gatherer of statistics, reported that in the first quarter of 2004 only 2% of the IT workers who lost their jobs did so due to offshoring.[10] The situation in Europe appeared to be similar with most job losses resulting from the economic recession – and not from offshoring. For example, The Netherlands had 15,000 unemployed IT workers in 2004 with the economic downturn responsible for the majority of these job losses.[11] Several studies looked a few years into the future and pointed out that the baby boom generation in North America and Western Europe is about to retire, which will create an enormous gap in jobs that are being conveniently filled by able engineers offshore.

Others pointed out that this was not the first time that US industries and US jobs were at stake from massive waves of offshoring. In fact such concerns have appeared roughly every decade: in the 1970s, it was the newly emergent Germany; in the 1980s, Japan; and in the 1990s, the fear of the North American Free Trade Agreement sending jobs to Mexico. Although, as noted previously, all these jobs were (generally) manufacturing or agricultural.

The consulting firm McKinsey & Company estimated that every dollar spent off-shore returns a gain for the US of 1.12 USD.[12] This is due to savings accrued to US corporations, to the value added from US labor redeployed, and from several other lesser items. The US technology industry association ITAA issued a report in 2004 project-ing that by 2008 more than 300,000 jobs will be created (not lost) due to offshoring.

A United Nations (UN) report on globalization of services also emphasized the positive.[13] It pointed to several factors that should benefit the wealthy nations: most offshoring of services still takes place among the wealthiest nations; that offshoring reduces costs, which benefits the home nations; that offshoring allows to shift resources to more productive activities; that revenues from offshoring in host nations is used, in part, for other imports from wealthy nations. The UN report did note that industrialized nations will have to manage the transitional loss of jobs through adjustment policies, such as training.

Companies' reactions

As the political debate swelled American companies were adjusting by offshoring under fire. Offshoring was not presenting a competitive threat to these companies. To the contrary, these companies perceived offshoring as vital to their competitiveness and opposed any protectionist obstacles. Both technology and non-technology compa-nies continued to expand their use of offshoring.

In spite of the political backlash, some corporate spokespeople were bold about offshoring. Carly Fiorina, superstar CEO of America's HP, famously pronounced that "There is no job that is America's God-given right anymore." Intel's CEO Craig Barrett was quoted repeatedly praising the benefits of offshoring.

Many corporations, though, were concerned about the public relations fallout. Some American and European companies chose to take the cautious political path by hold-ing IT employment at home steady, while expanding operations offshore. This way, they cannot be accused of displacing their employees at home with offshore workers.

The backlash caused most large US companies to enter into a covert offshore mode in which information was to be contained as much as possible. As an example, some IT managers were reluctant to speak "on the record" for this book because they were advised against doing so. Journalists were reporting similar experiences.

Some companies were careful to display public sensitivity to the job loss issue. For example, IBM declared publicly in 2004 that it would invest in retraining of those whose jobs were replaced due to offshoring. It set up a 25 million USD fund for retraining and gave employees on notice more time to find alternative positions within IBM. One 2004 survey[14] of American executives indicated a high degree of sensitivity to the social and economic consequences of offshoring: 40% supported a per head offshore tax for retraining, while 58% believed that firms that offshore should pay higher taxes.

It is more difficult to assess the overall impact on corporate decision-making. Namely, did the political backlash slow offshoring? Probably, not much. During the early 2000s, as the political debate emerged, offshoring was increasing very rapidly, probably as fast as the supplier nations of India, China, and others, could sustain. Polls in the US suggest that backlash had little impact: two American surveys in 2004 found that 86% and 72% of managers were *not* changing their plans due to backlash. A number of commentators began repeating the belief that the political backlash actually helped increase offshoring, since it focused so much attention on the subject that there could be no manager left in America who was not educated about the offshore option.

Reaction in India

With India being the largest beneficiary of offshoring, Indian companies and politicians were confused by the backlash. A common response was also articulated by then-Prime Minister Vajpayee who said "the very process of liberalization, on which we have been lectured for so many years, has created competitive skills ... we should not now drive a reverse process." India was disappointed that it had succeeded in creating a relatively open market economy, had succeeded through the grit of its people, and was now perceived as the bad guy. Others saw it as racism against brown-skinned people: when thousands of IT-related jobs moved to countries such as Ireland, this was never seen as a problem, was the refrain of some. India's software industry was careful not to stir the backlash: Indian firms stopped publicizing many of their new deals by 2003; and NASSCOM, the industry association, hired an American public relations firm.

In reality India had but limited exposure to many of the protectionist proposals that emerged in the US. Some proposed US laws called for restricting offshoring for tax-payer supported government projects. But India's business to the US government, or to any of the 50 states and many local governments, was well under 5% of its total business in 2003. India does, however, have a larger exposure in ITES due to data privacy regulations.

The longer-term policy issues

For advanced industrialized nations offshoring is not just an issue of job loss, as important as that is. It is also about loss of leadership in innovation. The Chinese and Indian technology centers are threatening to compete with the US and other leading technology nations in innovation activities.

Human capital and innovation

Once offshoring permeated the political landscape, Americans and Europeans began to look around and realize that a more serious threat than the direct loss of a few thousand software jobs may be the loss of innovative leadership. After all, technological

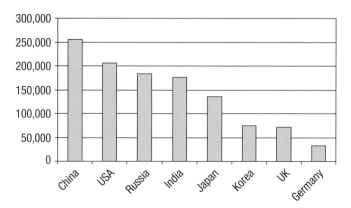

Figure 12.1 Number of degrees in Science and Engineering awarded in key nations (1999 or most recent year available).
Source: NSF Science and Engineering indicators.

innovation is the driver of economic growth and a key ingredient for a nation's long-term prosperity.

Innovation occurs because of many factors, but mostly due to a nation's human capital. In this regard, two trends became worrisome: fewer talented students were choosing careers in science and technology (S&T), and in the USA fewer talented foreign-born students were arriving for university degrees. The result has been a decline in the relative number of S&T-trained students in the USA relative to the global supply. Each of these trends is introduced below.

The educational ratios are beginning to favor Asian nations. Of university graduates about 5% of students in the US and Western Europe receive degrees in engineering compared to roughly 20% in Japan and roughly 40% in China. Figure 12.1 presents the number of students completing Science and Engineering degrees in selected countries indicating that the pipelines in key Asian nations are now quite robust. More recent estimates are less favorable to the US (focusing only on engineering), namely that the US graduated 60,000 engineers in 2002, while China graduated as many as 300,000 – more than four times more. The number of engineering degrees increased in China in 1995–1999 by 37% while they decreased in the US.

The number of Computer Science students began falling in the US. Computer Science and Computer Engineering students fell 23% in 2003.[15] In response, Bill Gates, Microsoft's Chairman and founder, embarked on a stumping tour at America's finest universities, the University of Illinois, Carnegie-Mellon, Cornell, MIT, and Harvard, to instill a sense of excitement in students about careers in computer science.[16]

For two centuries Germany has taken pride in its engineering and scientific innovations, which resulted in Nobel prizes, the invention of X-ray machines, and Mercedes Benz, among others. But Germany has seen a steadily declining number of students who choose to study engineering. Its technical universities are under-funded. The situation became so worrisome that Chancellor Schroeder declared 2004 as the "Year of Innovation."[17]

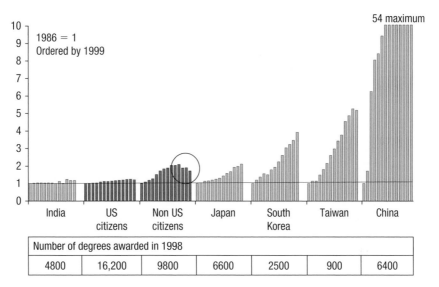

Figure 12.2 Growth in doctoral degrees awarded in 1986–1999 in key nations. *Source:* NSF Science and Engineering indicators, 2002.

The second worrisome trend is affecting the US. For a century American innovation was fueled in part by foreign-born S&T talent. In other words, the US has benefited from brain-drain in other nations. People in Silicon Valley jest that it was built on ICs which are Integrated Circuits, but also Indians and Chinese. But there are early indicators of a drop in those coming to the US and more of those who come return home to China, India, and Taiwan. The migration of S&T talent to the US has slowed since 2001,[18] partially because of increased industry opportunities in Asia, partially because of better education programs in Asia, and partially because of more difficult entry requirements since the September 2001 attack on the US.

Even the relatively favorable US position portrayed in Figure 12.1 is partially a result of a higher ratio of degrees given at the Masters and PhD level, many of which are awarded to foreign-born students, who are now returning home in greater numbers. Figure 12.2 shows the growth of doctoral degrees granted in the US versus key Asian nations. In the US nearly all of the growth can be explained by immigration of talented individuals from abroad. Again, it is likely that the data underestimate the actual contribution of immigrants since the data are for *US citizens* rather than *US-born* versus foreign-born. Among PhD holders working in US technology companies, half of the computer scientists were not US citizens.[19]

Techno-nationalism

Techno-nationalism is the notion that a nation's technological prowess will serve it well from a military-security perspective. The US, China, Russia, Israel, and Taiwan, are all countries with a strong techno-nationalist orientation.

Table 12.1 US rivalry with Japan in the 1980s versus China today.

Japan	China
High-value, high-wage advanced tech, "just like us"	Low-value, low-wage, advanced tech.
We have entrepreneurial advantage but they have industrial policy advantage	Entrepreneurial Using industrial policy
Rule of law	Limited rule of law
IP protections	IP theft model. 250 billion USD per year (US FBI est.)
Subsidized currency. Buying our debt	Subsidized currency. Buying our debt
National security – allies	National security – peer competitor

Source: Bonvillian, 2004[22]

Offshoring is viewed as a threat to US technology leadership. Besides the loss of jobs, offshoring is perceived as a threat to US national security: advanced technologies used by the military will be increasingly sourced from foreign nations; and, due to the porous nature of software, it is vulnerable to attacks and disruptions from insiders (programmers) in foreign nations working on the software that is offshored. US Senator Lieberman became the principal political proponent of this view arguing that "[offshoring] threatened to undermine America's innovation infrastructure."[20] Lieberman's science advisor positioned this techno-rivalry in Table 12.1. Not everyone in America subscribed to these techno-nationalist sentiments: "It is arrogant to think that we will be able to keep all research in the US," said the chairman of the Computer Research Association, an industry group.[21]

Concluding lessons

The social and economic consequences of software's migration offshore are still too young to assess. What is clear about this offshore migration is that it has no precedent in that, for the first time, jobs migrating offshore are of white collar, highly-educated professionals.

For the business decision-makers in America or Europe, the political backlash has resulted, in many cases, in the following reactions:

- Companies are keeping offshoring plans covert.
- Companies are expanding offshore while maintaining employment at home at steady levels.
- Companies replacing workers at home are making efforts to retrain targeted workers.

References

Aberdeen Group (2003). *Knowledge Transfer and On-site/Offshore Coordination are Key to the Success of Transco's Application Maintenance and Support Program*. Retrieved from http://www.aberdeen.com/2001/research/090318539.asp

AeA (2004). *Offshore Outsourcing in an Increasingly Competitive and Rapidly Changing World: A High-Tech Perspective*. Corporate Report.

Agrawal, V. & Farrell, D. (2004). Who wins in offshoring. *McKinsey Quarterly*, 4.

Allen, T. (1977). *Managing the Flow of Technology*. Cambridge, MA: MIT Press.

Ambler, S. W. (2002). Bridging the distance. *Software Development*, September. Retrieved from http://www.sdmagazine.com

Arora, A. & Athreye, S. (2001). The software industry and India's economic development. *United Nations University, Wider Discussion Paper*.

Aubert, B. A., Patry, M. & Rivard, S. (2002). Managing IT outsourcing risk: lessons learned. In: Hirschheim, R., Heinzl, A. & Dibbern, J. (eds), *Information Systems Outsourcing*. Berlin: Springer.

Bassellier, G., Reich, B. H. & Benbasat, I. (2001). Information technology competence of business managers: a definition and research model. *Journal of Management Information Systems*, 17(4), 159–182.

Bayman, S. (2002). *Remarks at the US–India Business Council Annual Meeting*, June 17.

Behrens, A. (2003). Brazilian software: the quest for an export-oriented business strategy. DRC Working Paper No. 21, London Business School.

Bennett, M. (1998). Intercultural communication: a current perspective. *Basic Concepts of Intercultural Communication: Selected Readings*. Intercultural Press.

Blackburn, R., Furst, S. & Rosen, B. (2003). Building a winning virtual team. In: Gibson, C. B. & Cohen, S. G. (eds), *Virtual Teams That Work*. San Francisco: Jossey-Bass.

Brockhoff, K. (1998). *Internationalization of Research and Development*. Springer.

Brown, C. & Magill, S. (1994). Alignment of the IS function with the enterprise: towards a model of antecedents. *MIS Quarterly*, 18(4), 371–403.

Bruell, N. (2003). Exporting software from Indonesia. *Electronic Journal on Information Systems in Developing Countries*, 13(7), 1–9.

Bulkeley, W. M. (2004). IBM documents give rare look at sensitive plans on offshoring. *The Wall Street Journal*, January 19.

Business Standard (2002). A backup base: GE taps critical mass with its new research lab in Bangalore, June 8.

Business Software Alliance (2003). *Eighth Annual BSA Global Software Piracy Study: Trends in Software Piracy 1994–2002*. Retrieved from http://www.bsa.org/globalstudy2003/pressreleases/loader.cfm?url=/commonspot/security/getfile.cfm&pageid=13034&hitboxdone=yes

Campoy, A. (2004). Think locally: Indian outsourcing companies have finally begun to crack the European market. *Wall Street Journal*, September 27.

Carmel, E. & Agarwal, R. (2002). The maturation of offshore sourcing of information technology work. *MIS Quarterly Executive*, 1(2).

Carmel, E. & Espinosa, A. (2004). *Online Programming Marketplaces*. Research Working Paper, Kogod School of Business, American University.

Carmel, E. (1999). *Global Software Teams: Collaborating Across Borders and Time Zones*. Prentice Hall-PTR.

Carmel, E. (2003). Taxonomy of new software exporting nations. *Electronic Journal on Information Systems in Developing Countries*, 13(2). Retrieved from http://www.is.cityu.edu.hk/research/ejisdc/vol13/v13r2.pdf

Carmel, E. (2003). The new software exporting nations: impacts on national well being resulting from their software exporting industries. *Electronic Journal on Information Systems in Developing Countries*, 13(3). http://www.is.cityu.edu.hk/research/ejisdc/vol13/v13r3.pdf

Carmel, E. (2003). The new software exporting nations: success factors. *Electronic Journal on Information Systems in Developing Countries*, 13(4). http://www.is.cityu.edu.hk/research/ejisdc/vol13/v13r4.pdf

Carr, S. (2004). IT salaries vary greatly by country. *Silicon.com*, June 7.

Chawla, S. (1998). Viability of offshore software development: a case study of international division of labor between Indian and US organizations. *Proceedings of the United Nations University Workshop Challenges and Opportunities for Globally Distributed Work*. Maastricht, November 24.

Chidamber, S. R. (2003). An analysis of Vietnam's ICT and software services sector. *Electronic Journal on Information Systems in Developing Countries*, 13(9), 1–11.

Chung, W. & Alcacer, J. (2003). Knowledge sources and foreign investment location in the US. *Conference of the Academy of International Business*, Monterrey, CA, June.

Chung, W. (2004). Companies that expand abroad: "knowledge seekers" vs. conquerors. *Knowledge@Wharton*, March 24. Retrieved from http://knowledge.wharton.upenn.edu/index.cfm?fa=viewArticle&id=952

CIO (2004). Ab ins Billige Ausland, in German, February 2.

Cisco Systems (2000). *Smart Valley Telecommuting Guide*. Retrieved from http://www.cisco.com/warp/public/779/smbiz/netsolutions/find/telecommuting/

Constantine, L. (1995). *Constantine on Peopleware*. Englewood Cliffs, NJ: Yourdon Press.

Conway, M. E. (1968). How do committees invent? *Datamation*, 14(4).

Cooper, C. (2004). Poll shows support for offshoring tax. *CNET News.com*, May 4, 2004.

Cooter, M. (2004). Companies underestimate cost of offshore outsourcing. *Computerweekly.com*, June 16.

Coward, C. T. (2003). Looking beyond India: factors that shape the global outsourcing decisions of small and medium sized companies in America. *Electronic Journal on Information Systems in Developing Countries*, 13(11), 1–2.

Cramton, C. (2003). Finding common ground in dispersed collaboration. *Organizational Dynamics*, 30(4), 356–367.

Cullen, S. & Willcocks, L. P. (2003). *Intelligent IT Outsourcing: Eight Building Blocks for Success*. Amsterdam: Elsevier.

Cummings, J. N., Wilson, J. M. & Pearce, B. M. (2002). Researching teams with multiple boundaries. *Proceedings of the 35th Hawaii International Conference on System Sciences* (HICSS 35). IEEE Press.

Cusumano, M., MacCormack, A., Kemerer, C. F. & Crandall, W. (2003). A global survey of software development practices. *MIT Sloan School Working Paper No. 178*, June.

Daft, R. L. & Lengel, R. H. (1986). Organizational information requirements, media richness and structural design. *Management Science*, 32(5), 554–571.

Davison, D. (2003). The top 10 risks of offshore outsourcing. *Meta Group Report*, November 14.

Davison, D. (2003). Offshore outsourcing subtleties. *Meta Group Report*, March 13.

Deloitte Consulting (2003). *Offshore Outsourcing. Is It the TCO Slasher It Promised to Be?* Biswas.

Deloitte Research (2004). *Making the Off-shore Call: The Road Map for Communications Operators*. Research Report.

Dennis, A. R. & Kinney, S. T. (1998). Testing media richness theory in the new media: the effects of cues, feedback, and task equivocality. *Information Systems Research*, 9(3), 256–274.

Desouza, K. C. (2003). Facilitating tacit knowledge exchange. *Communications of the Association for Computing Machinery*, 46(6), 85–88.

Diamondcluster (2004). *Global IT Outsourcing Study*. Research Report.

Dijk, M. P. Van & Wang, Q. (2003). The development of a software cluster in Nanjing. *Proceedings of the EADI Workshop on October 30–31*, Novara, Italy.

Dolven, B. (2004). China grooms global players. *Wall Street Journal*, February 25.

Dossani, R. & Kenney, M. (2003). *Went for Cost, Stayed for Quality? Moving the Back Office to India*. Working Paper, Stanford University Institute for International Studies.

Dutta, S., Lanvin, B. & Paua, F. (2003). *The Global Information Technology Report: Readiness for the Networked World*. Oxford University Press.

Economic Policy Institute (2004). *Offshoring, Frequently Asked Questions*. Retrieved June 2004 from http://www.epinet.org/content.cfm/issueguide_offshoring_faq

Economic Times (of India) (2003). It's patent mania for US companies in local units, December 17.

Economic Times (of India) (2004). Indians are faster, admits US firm, March 8.

Endsley, M. R. (1995). Toward a theory of situation awareness in dynamic systems. *Human Factors*, 37(1).

Espinosa, J. A. & Carmel, E. (2004). The impact of time separation on coordination in global software teams: a conceptual foundation. *Journal of Software Process Improvement and Practice* (8), 249–266.

Espinosa, J. A., Cummings, J. N., Wilson, J. M. & Pearce, B. M. (2003). Team boundary issues across multiple global firms. *Journal of Management Information Systems*, 19(4), 157–190.

Espinosa, J. A., Kraut, R. E., Lerch, F. J., Slaughter, S. A., Herbsleb, J. D. & Mockus, A. (2001). Shared mental models and coordination in large-scale, distributed software development. *International Conference in Information Systems*, New Orleans, LA.

EITO (2004). *European Information Technology Observatory 2004*, Frankfurt/Main, March.

Express Computer (2003). Europe beckons Indian software companies, April 21.

Feiman, J. (2004). *Economics of Application Development Outsourcing: Can Indiana Compete with India?* Slides presented at Gartner Symposium, March 28.

Fenema, P. C. van, & Qureshi, S. (2004). A phenomenological exploration of adaptation in a polycontextual work environment. *Proceedings of Hawaiian International Conference on Systems Sciences*.

Field, T. (2002). Man in the middle. *CIO Magazine*, April 1.

Fitzgerald, M. (2003). At risk offshore. *CIO Magazine*, November 15.

Florida, R. (2002). *The Rise of the Creative Class: and How Its Transforming Work, Leisure, Community and Everyday Life*. New York: Basic Books.

Forrester Research (2004). *Indian Offshore Suppliers, the Market Leaders*, April 7.

Gallaugher, J. & Stoller, S. (2004). Software outsourcing in Vietnam: a case study of a locally operating pioneer. *Electronic Journal of Information Systems in Developing Countries*, 17(1), 1–18.

Gengler, E. B. (2003). Ukraine and success criteria for the software exports industry. *The Electronic Journal of Information Systems in Developing Countries*, 13. Retrieved from http://www.ejisdc.org

GlobalSourcingNOW News (2004). SAP to Hire 1,900 in India, August 4.

Granstrand, O., Patel, P. & Pavitt, K. (1997). Multi-technology corporations: why they distributed rather than distinctive competencies. *California Management Review*, 39, 4.

Grinter, R. E., Herbsleb, J. D. & Perry, D. E. (1999). The geography of coordination: dealing with distance in R&D work. *Proceedings of International ACM SIGGROUP Conference on Supporting Group Work*, 306–315. Phoenix, Arizona: ACM Press.

Gumpert, D. E. (2003). US programmers at overseas salaries. *BusinessWeek Online*, December 2.

Hamm, S. (2004). Services. *BusinessWeek Online*, June 21.

Hampden-Turner, C. & Trompenaars, A. (1993). *The Seven Cultures of Capitalism*. New York: Currency Doubleday.

Harney, J. (2003). Cheaper, faster systems development using offshore outsourcing. *Outsourcing Journal*, July.

Hawk, S. & McHenry, W. (2005). The maturation of the Russian offshore software industry. *Journal of IT for Development*, 11(1).

Heeks, R. (1996). *India's Software Industry. State Policy, Liberalisation and Industrial Development*. Thousand Oaks, CA: Sage Publications.

Heeks, R. B. (1999). Software strategies in developing countries. *Communications of the ACM*, June, 42(6), 15–20.

Herbsleb, J. D. (2003). Keynote presentation. *The International Workshop in Global Software Development, International Conference on Software Engineering*.

Herbsleb, J. D. & Mockus, A. (2003). An empirical study of speed and communication in globally-distributed software development. *IEEE Transactions on Software Engineering*, 29(3), 1–14.

Hinds, P. & Weisband, S. (2003). Knowledge sharing and shared understanding in virtual teams. In: Gibson, C. B. & Cohen, S. G. (eds), *Virtual Teams That Work*. San Francisco: Jossey-Bass.

Hindu Business Line (2002). India's treasure is its intellectual capital, October 7.

Hira, R. (2003). *On the Offshoring of High-Skilled Jobs*. Testimony to the US House of Representatives Committee on Small Business, October 20.

Hiskisson, R. E., Hitt, M. A. & Ireland, R. D. (2004). *Competing for Advantage*. Ohio: Thomson/ Southwestern.

Hofstede, G. (1991). *Cultures and Organizations: Software of the Mind*. London: McGraw Hill.

Hofstede, G. (1993). Cultural constraints in management theories. *Academy of Management Executive*, 7(1), 81–93.

InfoTech (1992). International studies of software and related services. India's software and services export potential and strategies. *The World Bank-Funded Report for the Department of Electronics Government of India. Volumes I and II*. New Jersey: InfoTech Consulting Inc.

Intel (2003). *YourTime: Email Effectiveness Program*. Retrieved from https://www.itsharenet.org/ kshowcase/view/view_item?item_key=49909e9b26ee08341685fd7f4fb5e9888d59254b

International Trade Centre (2002). *Country Profile: Lithuania*. Geneva.

Irwin, B. P. (2001). Offshore corporations: a brief introduction. *Harvard Business School Case 9-799-119*.

Jackson, T. W., Dawson, R. & Wilson, S. (2003). Understanding email interaction increases organizational productivity. *Communications of the ACM*, August.

Jarvenpaa, S. & Leidner, D. (1999). Communication and trust in global virtual teams. *Organization Science*, Winter, 791–815.

Jensen, M. (2001). Afriboxes, telecenters, cybercafes: ICT in Africa. *UN-TCDC: Cooperation South Journal*, 1, 97–109.

Johansson, C., Dittrich, Y. & Juustila, A. (1999). *Software Engineering Across Boundaries – Student Project in Distributed Collaboration*. Working Paper, University of Karlskrona, Sweden.

Jones, C. (1994). Globalization of software supply and demand. *IEEE Software*, November, 11(6), 17–24.

Joseph, K. J. (2002). Growth of ICT and ICT for development. Realities of the myths of the Indian experience. *United Nations University, WIDER Discussion Paper No. 2002/78*.

Jurison, J. (2002). Applying traditional risk-return analysis to strategic IT outsourcing decisions. In: Hirschheim, R., Heinzl, A. & Dibbern, J. (eds), *Information Systems Outsourcing*. Berlin: Springer.

Kane, M. (2003). Cisco sues Huawei over patents. *Cnet News*, January 23. Retrieved from http://Business2-cnet.com.com/2102-1033-981811.html

Karnitschnig, M. (2004). Vaunted German engineers face competition from China. *New York Times*, July 15.

Kearney, A. T. (2004). *Selecting IT Activities for Offshore Locations*. Research Report.

Kearney, A. T. (2004). *Making Offshore Decisions*. Research Report.

Kessler, M. (2004). Fewer college students choose computer majors. *USA Today*, August 8.

Kiel, L. K. (2003). Experiences in distributed development: a case study. *Proceedings of the International Workshop in Global Software Development, International Conference on Software Engineering*.

Kiesler, S. & Cummings, J. N. (2002). What do we know about proximity and distance in work groups? A legacy of research. In: Hinds, P. & Kiesler, S. (eds), *Distributed Work*. MIT Press.

Kirkpatrick, K. E. (2004). Taking advantage of IM. *Forbes.com*, April 9.

Knoll, K. & Jarvenpaa, S. L. (1995). Learning to work in distributed global teams. *Proceedings of the 28th Annual Hawaii International Conference on Systems Science*, 92–101.

Koch, C. (2004). Bursting the CMM hype. *CIO Magazine*, March 1.

Kock, N. (2004). The psycho-biological model: toward a new theory of computer-mediated communication based on Darwinian evolution. *Organization Science*, 15(3).

Kogut, B. & Singh, H. (1988). The effect of national culture on the choice of entry mode. *Journal of International Business Studies*, 23(Spring), 29–53.

Konrad, R. (2003). Offshoring dulls startups' US presence. *InformationWeek*, December 15.

Kraemer, K. & Dedrick, J. (1999). *National Policies for the Information Age: IT and Economic Development*. Center for Research on Information Technology and Organizations, University of California.

Kripalani, M. & Engardio, P. (2003). The rise of India. *BusinessWeek*, December 8.

Krishnamurthy, S. (2002). Cave or community? An empirical examination of 100 mature open source projects. *First Monday*, 7(6). Retrieved from http://www.firstmonday.org/issues/issue7_6/krishnamurthy/

Krishnan, R. T. & Prabhu, G. N. (2004). Software product development in India: lessons from six cases. In: D'Costa, A. P. & Sridharan, E. (eds), *India in the Global Software Industry*. New York: Palgrave Macmillan.

Kuemmerle, W. (1997). Building R&D capabilities abroad. *Harvard Business Review*, March–April.

LaFave, R. (2004). Career watch. *Computerworld*, July 5.

Lanvin, B. & Quian, C. Z. (2004). Poverty e-readication: using ICT to meet millennium development goals. In: Dutta, S., Lancin, B. & Paua, F. (eds), *The Global Information Technology Report 2003–2004*. Oxford University Press.

Laudon, K. C. & Laudon, J. (2003). *Management Information Systems*. Pearson.

Lee, J. N., Miranda, S. & Kim, Y. (2004). IT outsourcing strategies: universalistic, contingency, and configurational explanations of success, *Information Systems Research*, 15(2).

Lieberman, J. (2004). *Lieberman Calls Offshore Outsourcing of US Jobs Tip of Economic Iceberg*. Press Release, US Senate, May 11.

Lieberman, J. (2004). *Offshore outsourcing and America's competitive edge: losing out in the high technology R&D and services sectors*. Office of US Senator Lieberman, May 11.

Lohr, S. (2004). Microsoft, amid dwindling interest, talks up computing as a career. *New York Times*, March 1.

Lu, M., Wynn, E., Watson-Manheim, B. & Chudoba, K. (2003). Understanding virtuality in a global organization: toward a virtuality index. *Proceedings of 24th International Conference in Information Systems*.

Majchrzak, A. & Malhotra, A. (2003). *Deploying Far Flung Teams: A Guidebook for Managers.* Society for Information Management, May.

Mark, G. (2002). Conventions for coordinating electronic distributed work: a longitudinal study of groupware use. In: Hinds, P. & Kiesler, S. (eds), *Distributed Work*. MIT Press.

Martin, K. (2003). Research and development goes offshore. *Meta Group News Analysis*, September 12.

Massey, A. P., Montoya-Weiss, M., Hung, C. & Ramesh, V. (2001). Global virtual teams: cultural perceptions of task–technology fit. *Communications of the ACM*, 44(12), 83–84.

Maznevski, M. L. & Chudoba, K. M. (2000). Bridging space over time: global virtual team dynamics and effectiveness. *Organization Science*, 11(15), 473–492.

McCarthy, J. C. (2002). 3.3 million US Service Jobs To Go Offshore. *Forrester News brief*, November 11. Retrieved from http://www.forrester.com

McCarthy, J. C. (2003). Unlocking the savings in offshore. *Forrester Research Report*, February 3.

McCarthy, J. C. (2003). Users offshore evolution and its governance impact. *Forrester Brief*, December 4.

McDougall, P. (2003). Opportunity on the line. *InformationWeek*, October 20.

McDougall, P. (2004). GE expects to send more work offshore. *InformationWeek*, May 17.

McKinsey & Co (1999). *The Indian IT Strategy*. New York: McKinsey & Co; New Delhi: NASSCOM.

Meta Group (2004). *METAspectrum Report on the Offshore Outsourcing Market*, October.

Mintzberg, H. (1993). *Structure in Fives: Designing Effective Organizations*. Prentice Hall: Englewood Cliffs, New Jersey.

Mishra, P. (2002). GE changes outsourcing paradigm for India. *Express Computer*, April 15.

Mitter, S. (2004). *Offshore Outsourcing of Information Processing Work and Economic Empowerment of Women*. Presentation at the World Bank, Washington D.C., June 2.

Moore, S. & Martorelli, W. (2004). Indian offshore suppliers: the market leaders. *Forrester Market Overview*, April 7.

Morales, A. W. (2004). Has this trend sprung a leak? *Software Development Magazine*, January. Retrieved from http://www.sdmagazine.com/

Morris, M. W. (1999). *Email and the Schmooze Factor*. Research Summary. Retrieved from http://www.gsb.stanford.edu/research/reports/1999/morris.html

Murphy, C. (2003). Future view: software jobs will be mechanized in the long run. *InformationWeek*, November 17.

Narula, R. (2001). Choosing between internal and non-internal R&D activities: some technological and economic factors. *Technology Analysis and Strategic Management*, 13(3).

Nicholson, B. & Sahay, S. (2003). Building Iran's software industry: as assessment of plans and prospects. *Electronic Journal on Information Systems in Developing Countries*, 13(6). http://www.is.cityu.edu.hk/research/ejisdc/vol13/v13r6.pdf

Nicholson, B. & Sahay, S. (2004). Embedded knowledge and offshore software development. *Information and Organization*, October, 329–365.

Niederman, F. (2004). IT employment prospects in 2004: a mixed bag. *IEEE Computer*, January.

Nisbett, R. E. (2003). *The Geography of Thought: How Asians and Westerners Think Differently and Why*. New York: The Free Press.

Overby, S. (2002). A buyer's guide to offshore outsourcing. *CIO Magazine*, November 15.

Overby, S. (2003). The hidden costs of offshore outsourcing. *CIO Magazine*, September 1.

Ovum Holway (2003). *The Offshore Services Report 2003*.

Parnas, D. L. (1972). On the criteria to be used in decomposing systems into modules. *Communications of the ACM*, 15(12).

Parthasarathi, A. & Joseph, K. J. (2004). Innovation under export orientation. In: D'Costa, A. P. & Sridharan, E. (eds), *India in the Global Software Industry*. New York: Palgrave Macmillan.

Peters, P. & Dulk, L. den (2003). Cross cultural differences in managers' support for home-based telework: a theoretical elaboration. *International Journal of Cross Cultural Management*, 3(3), 329.

Philips, R. A. (1998). *Guide to Software Export. A Handbook for International Software Sales*. New York: The International Business Press.

Prism Economics and Analysis (2004). *Trends in the Offshoring of IT Jobs*, April.

Reichgelt, H. (2000). Software engineering services for export and small developing countries. *Information Technology for Development*, 9(2), 77–90.

Resnick, P. (2002). Beyond bowling together: socio-technical capital. In: Carroll, J. M. (ed.), *Human–Computer Interaction in the New Millennium*. Boston: Addison-Wesley.

Ricciuti, M., Frauenheim, E. & Yamamoto, M. (2004). The next battlefields of advanced technology. *CNET News.com*, May 7.

Roberts, B. (2004). The perfect storm brews offshore. *Electronic Business*, March 1.

Roberts, J. (2000). From know-how to show-how? Questioning the role of ICT in knowledge transfer. *Technology Analysis and Strategic Management*, 12(4).

Rubin, H. A. & Jaramillo, P. (2004). *Outsourcing: An Analysis of the Current State of Offshore Outsourcing in New York City Based Companies*. Retrieved from http://www.newjobsforny.org/OutsourcingReport.php

Sahay, S., Nicholson, B. & Krishna, S. (2003). *Global IT Outsourcing: Software Development Across Borders*. Cambridge: Cambridge University Press.

Sambamurthy, V. & Zmud, R. (1999). Arrangements for information technology governance: a theory of multiple contingencies. *MIS Quarterly*, 23(2), 261–290.

SandHill Group (2003). *The Roadmap to Offshore Success*. Corporate Report, August.

Saxenian, A. (2003). *Government and Guanxi: China's Software Industry in Transition*. University of California Working Paper. Retrieved from http://www.sims.berkeley.edu/~anno/papers/softwareinchina.pdf

Saxenian, A. (2004). The Silicon Valley connection: transnational networks and regional development in Taiwan, China and India. In: D'Costa, A. P. & Sridharan, E. (eds), *India in the Global Software Industry*. New York: Palgrave Macmillan.

Scacchi, W. (2002). Understanding the requirements for developing open source software systems. *IEE-Software Proceedings*, February.

Schneider, S. C. & Barsoux, J. (1993). *Managing Across Cultures*. London: Prentice Hall.

Sengupta, S., Gupta, I.& Singh, S. (2004). GECIS: The house that Jack built … and Jeffrey is about to sell. *Businessworld*, October 11.

Shewell, C. (2000). *Good Business Communication Across Cultures*. Briston, UK: Mastek.

Sridharan, E. (2004). Evolving towards innovation? The recent evolution and future trajectory of the Indian software industry. In: D'Costa, A. P. & Sridharan, E. (eds), *India in the Global Software Industry*. New York: Palgrave Macmillan.

Standish Group (2003). *CHAOS Chronicles*. Research Report.

Steinfield, C., Jang, C. & Pfaff, B. (1999). Supporting virtual team collaboration. *Proceedings of Group '99*. Available from the ACM.

Strassman, P. A. (2004). Most outsourcing is still for losers. *Computerworld*, February 2.

Survey of Computer Jobs in Israel (2004). Retrieved from http://www.cji.co.il

Swigger, K., Alpaslan, F., Brazile, R. & Monticino, M. (2004). Effects of culture on computer-supported international collaborations. *International Journal of Human–Computer Studies*, 60, 365–380.

Tedesco, J. (2004). Bucking the offshore trend. *Computerworld*, June 17.

Teicher, S. A. (2004). A not so simple path. *Christian Science Monitor*, February 23.

Terdiman, R. (2001). CIO update: a world of choices for application outsourcing. *Gartner InSide*, December 5.

The Economist (2004). More gain than pain: why America wins but Germany loses, July 17.

The Guardian (2003). Xansa retreats from the continent, December 5.

Thurm, S. (2004). Lesson in India, not every job translates overseas. *Wall Street Journal*, March 3.

Tidd, J. & Trewhella, M. (1997). Organizational and technological antecedents for knowledge creation and learning. *R&D Management*, 27, 359–375.

Tjia, P. (1999). *Market Survey – Computer Software and IT Services from Developing Countries*. CBI Rotterdam.

Tjia, P. (2003). The software industry in Bangladesh and its links to The Netherlands. *Electronic Journal on Information Systems in Developing Countries*, 13(5), 1–8.

Tjia, P. (2004). *Offshore Outsourcing*. GPI Consultancy, Markt Rapport, in Dutch, February.

Trompenaars, F. & Woolliams, P. (2004). *Business Across Cultures*. Capstone.

US Department of Labor, Bureau of Labor Statistics (2004). *Extended Mass Layoffs Associated with Domestics and Overseas Relocations, First Quarter 2004*. News Release, June 10.

United Nations Conference on Trade and Development (2002). *World Investment Report 2002: Transnational Corporations and Export Competitiveness*. New York: United Nations Publications. Retrieved from http://www.unctad.org/Templates/webflyer.asp?docid=2574&intItemID=2770&lang=1

United Nations Conference on Trade and Development (2004). *World Investment Report 2004: The Shift to Services*. New York: United Nations Publications. Retrieved from http://www.unctad.org/Templates/webflyer.asp?docid=5209&intItemID=1397&lang=1

Vernon, R. (1966). International investment and international trade in the product cycle. *Quarterly Journal of Economics*, May.

Vietnam Economic News (2004). IT revenue exceeds goal, July 28.

Waters, R. (2003). It has $3.3 billion to buy IT and GE still wants more for less. *Financial Times*, April 30.

Weinstein, L. (2004). Outsourced and out of control. *Communications of the ACM*, February.

Wells, J. (1998). *IT Services. A Handbook for Exporters from Developing Countries*. International Trade Centre UNCTAD/WTO, Geneva.

Wells, L. (1968). A product life cycle for international trade? *Journal of Marketing*, July.

World Markets Research Centre (2003). *Country Risk Reports*, February.

Yamamoto, M. & Said, K. (2004). Offshoring. *News.com*, May 4.

Yee, C. M. (2000). Let's make a deal. *Wall Street Journal*, September 25.

End notes

Preface

1. For the use of the term "offshore" for tax havens see: Irwin, B. P. (2001). Offshore corporations: a brief introduction, *Harvard Business School case 9-799-119*.
2. Nearsourcing is found in Sahay, S., Nicholson, B. & Krishna, S. (2003). *Global IT Outsourcing: Software Development Across Borders*. Cambridge University Press.

Chapter 1

1. G7 nations are the major economic powers of the post World War II period: USA, Canada, Britain, France, Germany, Italy, and Japan.
2. The International product cycle was originally applied to explain foreign direct investment in manufacturing. Vernon, R. (1966). International investment and international trade in the product cycle. *Quarterly Journal of Economics*, May.
3. United Nations Conference on Trade and Development (2004). *World Investment Report 2004: The Shift to Services*. www.unctad.org
4. Interview with Simonyi in: Murphy, C. (2003). Future view: software jobs will be mechanized in the long run. *InformationWeek*, November 17.
5. Carmel, E. & Agarwal, R. (2002). The maturation of offshore sourcing of information technology work. *MIS Quarterly Executive*, 1(2). The stage model was also adapted in: McCarthy, J. C. (2003). Users offshore evolution and its governance impact. *Forrester Brief*, December 4.
6. Diamondcluster (2004). *Global IT Outsourcing Study*. Research Report.
7. This composite cost saving is based on studies by Deloitte Consulting, Sand Hill Group, and Gartner. The latter two used only US firms. The results are reported in greater detail in Chapter 2, Offshore Economics and Offshore Risks.
8. Based on presentation slides of Stan Lepeak of the Meta Group, presented at *Russian Outsourcing Conference* in St. Petersburg, June 2004.
9. McCarthy, J. C. (2003). Users offshore evolution and its governance impact. *Forrester Brief*, December 4.
10. Chung, W. & Alcacer, J. (2003). Knowledge sources and foreign investment location in the US. *Conference of the Academy of International Business*, Monterrey, CA; June.
11. Another important strategic approach in software R&D is through acquisitions targeted at gaining talent for innovation. This is practiced by global technology firms that buy technology companies in Israel. More than one hundred Israeli firms, from start-ups to established companies, were acquired by foreign technology companies in the period from the mid 1990s through the early 2000s.
12. Carmel, E. (1999). *Global Software Team: Collaborating Across Borders and Time Zones*. Prentice Hall-PTR.

13. Morales, A. W. (2004). Has this trend sprung a leak? *Software Development Magazine*, January. www.sdmagazine.com

14. Morales, A. W. (2004). *Ibid.*

15. The estimate appears in: Martin, K. (2003). Research and development goes offshore. *Meta Group News Analysis*, September 12.

16. Forrester (2003). *Unlocking the Savings in Offshore.* Research Report, February.

17. Koch, C. (2004). Bursting the CMM Hype, *CIO Magazine*, March 1.

18. Cusumano, M., MacCormack, A., Kemerer, C.F. & Crandall, W. (2003). A global survey of software development practices. *MIT Sloan School Working Paper* No. 178, June.

19. Study by Administrative Staff College of India as appeared in: Yamamoto, M. & Said, K. (2004). Offshoring. *News.com*, May 4.

20. Lieberman, J. I. (2004). *Offshore Outsourcing and America's Competitive Edge: Losing Out in the High Technology R&D and Services Sectors.* Office of US Senator Lieberman, May 11.

21. Deloitte Research compiled an estimate that gives us some idea of the magnitudes involved. Deloitte estimated that by 2009, the global financial services industry will shift two million jobs offshore, while in the communications industry 275,000 jobs will be shifted offshore. Some of these jobs are in software, while the majority are in IT-enabled services, such as call centers. Deloitte Research (2004). *Making the Off-shore Call: The Road Map for Communications Operators.* Research Report.

22. Roberts, B. (2004). The perfect storm brews offshore. *Electronic Business*, March 1.

23. Teicher, S. A. (2004). A not so simple path. *Christian Science Monitor*, February 23.

24. Express Computer (2003). *Europe Beckons Indian Software Companies*, April 21.

25. EITO (2004). *European Information Technology Observatory 2004*, Frankfurt/Main, March.

26. Ovum Holway (2003). *The Offshore Services Report.* Ovum Holway underlined the relatively small size of the sector: if offshore were a single player, it would rank at number 13 in Ovum Holway's provisional rankings of the top 40 suppliers of software and IT services in the UK market. Nevertheless, it is estimated that by 2006, some 20,000–25,000 IT jobs would be lost offshore. Another report, from Evalueserve forecasted that 250,000 UK jobs (including IT) would move offshore by 2010. Evalueserve (2003). The *Impact of Global Sourcing on the UK Economy 2003–2010.*

27. Roland Berger/UNCTAD data that appears in: Campoy, A. (2004). Think locally: Indian outsourcing companies have finally begun to crack the European market. *Wall Street Journal*, September 27.

28. *GlobalSourcingNOW News* (2004). SAP to Hire 1900 in India, August 4.

29. *Economist* (2004). More gain than pain: why America wins but Germany loses, July 17.

30. *CIO* (2004). Ab ins Billige Ausland, in German, February 2.

31. Tjia, P. (2004). *Offshore Outsourcing.* GPI Consultancy, Markt Rapport, in Dutch, February.

32. Number of US software product firms is based on zapdata.com

33. Number of US IT services firms is a rough estimate by Specifics Analytics.

34. Bureau of Labor Statistics, US Department of Labor (2004). *Occupational Outlook Handbook*, 2004–2005 Edition.

35. IDC statistics as appeared in: Hamm, S. (2004). Services. *BusinessWeek Online*, June 21.

36. *Economic Times* (of India) (2004). *Indians are Faster, Admits US Firm*, March 8.

37. Ovum Holway (2003). *The Offshore Services Report.*

38. McDougall, P. (2003). Opportunity on the line. *InformationWeek*, October 20.

39. Examples are found in: *Prism Economics and Analysis* (2004). Trends in the Offshoring of IT Jobs, April; and Dossani, R. & Kenney, M. (2003). *Went for Cost, Stayed for Quality? Moving the Back Office to India.* Working Paper, Stanford University Institute for International Studies.

40. Jensen, M. (2001). Afriboxes, Telecenters, Cybercafes: ICT in Africa. In: UN-TCDC: *Cooperation South Journal*, 1, 97–109.

41. Forrester (2002). *3.3 Million US Service Jobs to Go Offshore*, Research Brief, November 11.

Chapter 2

1. Harney, J. (2003). Cheaper, faster systems development using offshore outsourcing. *Outsourcing Journal*, July.

2. Konrad, R. (2003). Offshoring dulls startups US presence. *InformationWeek*, December 15.

3. Tjia, P. (2004). *Offshore Outsourcing*. GPI Consultancy, Markt Rapport, in Dutch, February.

4. Slightly different rankings at the bottom of the wage scale are found in a 2004 study by Mercer Human Resource Consulting: the lowest wage rates are in the Philippines, followed by Vietnam, Bulgaria, Malaysia, Indonesia, and India.

5. Generally, many of these sources were compiled when the US dollar was relatively strong. All numbers not marked by a letter are based on authors' notes. (a) IT toolbox survey, 2003 in Niederman, F. IT Employment prospects in 2004: a mixed bag. *IEEE Computer*, January 2004; (b) Gengler, E. B. (2003). Ukraine and success criteria for the software exports industry, electronic. *The Electronic Journal of Information Systems in Developing Countries*, 13, www.ejisdc.org; (c) Computer jobs in Israel (2004) *Survey of Computer Jobs in Israel*, programmer with 3+ years of experience, www.cji.co.il; (d) Aberdeen 2001 data appearing in Field, T. Man in the middle, *CIO Magazine*, April 1, 2002; (e) Silicon.com data in Carr, S. (2004). IT salaries vary greatly by country. Silicon.com, June 7; (f) Overby, S. (2002). A buyers guide to offshore outsourcing, *CIO Magazine*, November 15; (h) Meta Group data received by authors June 2004.

6. Ratios use data provided by Value Leadership Group, 2004.

7. 2004 Survey of Computer Jobs in Israel, www.cji.co.il

8. Bulkeley, W. M. (2004). IBM documents give rare look at sensitive plans on offshoring. *The Wall Street Journal*, January 19.

9. Online marketplaces include: Rent A Coder, ELance, and ScriptLance.

10. Carmel, E. & Espinosa, A. (2004). *Online Programming Marketplaces*; Research Working Paper, Kogod School of Business, American University.

11. Transactions Costs is an economic concept introduced by the economist Ronald Coase, in a 1937 essay titled "The Nature of the Firm." Coase was later awarded the Nobel Prize in economics, in part for his contribution on transaction costs in this essay.

12. Thurm, S. (2004). Lesson in India, not every job translates overseas. *Wall Street Journal*, March 3.

13. Overby, S. (2003). The hidden costs of offshore outsourcing, *CIO Magazine*, September 1.

14. Feiman, J. (2004). *Economics of Application Development Outsourcing: Can Indiana Compete with India?* Gartner Symposium, March 28, presentation slides.

15. Computation for extended onsite work includes charge rate, plus living expenses, plus travel.

16. Davison, D. (2003). Offshore outsourcing subtleties. *Meta Group Report*, March 13.

17. Overby, S. (2003). The hidden costs of offshore outsourcing, *CIO Magazine*, September 1.

18. Search & Contract costs, in percent, tend to be higher for small organizations, and lower for very large contracts.

19. Deloitte Research (2004). *Making the Off-shore Call: The Road Map for Communications Operators*. Research Report.

20. Roberts, B. (2004). The perfect storm brews offshore. *Electronic Business*, March 1.

21. Rubin, H. A. & Jaramillo, P. (2004). Outsourcing: an analysis of the current state of offshore outsourcing in New York City based companies. www.newjobsforny.org/ OutsourcingReport.php

22. Cooter, M. (2004). Companies underestimate cost of offshore outsourcing. *Computerweekly.com*, June 16.

23. In the 2003 Standish study: project success rates were 34%. Project failures were 15% of all projects. Challenged projects account for the remaining 51%. Standish Group (2003). CHAOS Chronicles, company research report, www.standishgroup.com

24. Country risk, termed "Political risk" is insurable by some major insurance companies as well as by OPIC (for US investors).

25. The discussion of risks is based in part on Meta Group data in: Davidson, D. (2003). The top 10 risks of offshore outsourcing. *Meta Group Report*, November 14.

26. Other risks in outsourcing include lock-in, service debasement, and costly contracting amendments. General outsourcing risks are discussed in two sources: Aubert, B. A., Patry, M. & Rivard, S. (2002). Managing IT outsourcing risk: lessons learned. In: Hirschheim, R., Heinzl, A. & Dibbern, J. (eds), *Information Systems Outsourcing*. Berlin: Springer. Jurison, J. (2002). Applying traditional risk-return analysis to strategic IT outsourcing decisions. In: Hirschheim, R., Heinzl, A. & Dibbern, J. (eds), *Information Systems Outsourcing*. Berlin: Springer.

27. Kane, M. (2003). Cisco sues Huawei over patents. *CNET News*, January 23. Business2-cnet.com/2102-1033-981811.html

28. Fitzgerald, M. (2003). At risk offshore, *CIO Magazine*, November 15.

29. Weinstein, L. (2004). Outsourced and out of control. *Communications of the ACM*, February.

30. Tedesco, J. (2004). Bucking the offshore trend. *Computerworld*, June 17.

31. SandHill Group (2003). *The Roadmap to Offshore Success*. Corporate Report.

Chapter 3

1. A. T. Kearney, (2004). *Selecting IT Activities for Offshore Locations*, Corporate Research Report.

2. Detailed methodologies for the acquisition of IT services can be useful. A European example of a best practice library is Information Services Procurement Library (ISPL), which describes the process to acquire external IT services in detail.

3. Forrester Research (2003). *Unlocking the Savings in Offshore*, February.

4. Consulting and research firms that publish reports on offshoring include: Aberdeen Group, Evalueserve, Forrester, Gartner, GPI Consultancy, IDC, Meta Group and Ovum Holway. These organizations publish reports for free or for a fee.

5. Consultancies specializing in offshoring include: GPI Consultancy, Morgan Chambers, neoIT, Orbys, PA Consulting and TPI.

6. An example is Offshore Development Group (ODG), which conducts consultant verifications.

7. In most situations, organizations decide to outsource work to just one provider. Working with a single provider avoids procurement time and costs. It also enhances knowledge transfer. Some, typically larger clients choose a multi-provider strategy, and work with several "preferred suppliers", or have one leading provider and also a mid-sized provider. Working with several providers keeps the competitive spark and results in greater flexibility in terms, skills, and manpower scheduling. Multiple partners will also diversify the risks in case of non-performance. In general, however, more than three or four partnerships are difficult to manage – even for the largest customers.

8. The online marketplaces are like an *eBay* for buyers and sellers of software services. Examples include: Elance, ProjectPool, Rent A Coder, Scriptlance, and Smarterwork.

9. Coward, C. T. (2003). Looking beyond India: factors that shape the global outsourcing decisions of small- and medium-sized companies in America. *Electronic Journal on Information Systems in Developing Countries*, 13(11), 1–2.

10. Diamondcluster (2004). *Global IT Outsourcing Study*, Report.

11. Chawla, S. (1998). Viability of offshore software development: a case study of international division of labor between Indian and US organizations. *The United Nations University Workshop 'Challenges and Opportunities for Globally Distributed Work'* Maastricht, November 24.

12. Deloitte Consulting (2003). *Offshore Outsourcing. Is It The TCO Slasher It Promised To Be?* Biswas.

Chapter 4

1. This phenomenon, of the globalization of software development, has been of interest since the 1990s. Jones, C. (1994). Globalization of software supply and demand. *IEEE Software*, 11(6), November, 17–24. Heeks, R. B. (1999). Software strategies in developing countries. *Communications of the ACM*, 42(6), 15–20.

2. G7 countries are: USA, Canada, UK, France, Germany, France, Italy, and Japan.

3. Other advanced economies' industries have also had moderate success: Australia, Spain, Belgium, and the other Nordic countries.

4. This section and the tiered taxonomy is adapted from: Carmel, E. (2003). Taxonomy of new software exporting nations. *The Electronic Journal on Information Systems in Developing Countries*, 13(2), May.

5. Software export revenues should be treated with great caution. While India's software association, NASSCOM, is considered reliable in its estimates for India's software sector, in many nations these data are either difficult to find or unavailable. Worse, these data are prone to exaggeration; it is in the interest of all national parties – governments, industry associations, foreign consultants – to inflate the export revenues.

6. This discussion of offshore location factors has its origins in the domain of international R&D. It is partially based on: Brockhoff, K. (1998). *Internationalization of Research and Development*. Springer.

7. Kogut, B. & Singh, H. (1988). The effect of national culture on the choice of entry mode. *Journal of International Business Studies*, 23(Spring), 29–53.

8. A further refinement on the *location decision* distinguishes between these two types of offshore sites: the first is augmentation, where the site is located next to scientific excellence and the direction of tech transfer is back to home. The second is Exploitation, where the site is next to existing manufacturing/marketing, and the direction of tech trans is from home to foreign. Kuemmerle, W. (1997). Building R&D capabilities aboard. *Harvard Business Review*, March–April.

9. United Nations Conference on Trade and Development (2002). *World Investment Report 2002: Transnational Corporations and Export Competitiveness*. New York: United Nations Publications. Retrieved from http://www.unctad.org/Templates/webflyer.asp? docid=2574& intItemID=2770&lang=1

10. Kearney, A. T. (2004). Making Offshore Decisions. Research Report.

11. Some country risks are lower offshore: the country rated the fourth highest risk of terrorism is not an obscure offshore destination, but the United States, according to Global Insight (with the UK as the highest in Europe).

12. Dutta, S., Lanvin, B. & Paua, F. (2003). *The Global Information Technology Report: Readiness for the Networked World*. Published by Oxford University Press.
13. Dutta, *et al. Ibid.*
14. The benefits were culled from a various sources including embassies, as well as: Yee, C. M. (2000). Lets make a deal. *Wall Street Journal*, September 25.
15. Unified Modeling Language, a design language and modeling technique used worldwide.
16. Study by Administrative Staff College of India as appeared in Yamamoto and Said 2004. *Ibid.*
17. A number of major Japanese firms do software work in China including: NTT Data Corporation, NEC Soft, and Hitachi Software Engineering.
18. Presentation by Denis Simon of Rensselaer Polytechnic University, AAAS, April 2004.
19. Dolven, B. (2004). China grooms global players. *Wall Street Journal*, February 25.
20. Adapted from: Saxenian, A. (2003). *Government and Guanxi: China's Software Industry in Transition*, University of California Working Paper.
21. Market access is the demand by the host country to invest in human and fixed infrastructure in exchange for foreign direct investment.
22. Hawk, S. & McHenry, W. (2005). The maturation of the Russian offshore software industry. *Journal of IT for Development*.
23. Hawk, S. & McHenry, W. *Ibid.*
24. Bardhan, A. & Kroll, C. (2004). Research presented at *Russian Outsourcing Software Summit*, St. Petersburg.
25. Auriga presentation at the *2004 Russian Outsourcing Software Summit*, St. Petersburg.
26. All employment figures for Israel are for 2004.
27. Information on Latvia relied in part on information provided by Jânis Iesalnieks of DATI Deutschland.
28. Romanian firms are also among the most active bidders in online programming marketplaces.
29. *Country Risk Reports*, compiled by World Markets Research Centre, February 2003.
30. Information on Vietnam was based on: Duong, N. T. (2004). Software industry development in Vietnam', presented at the *IIPI Conference Strategies for Building Software Industries in Developing Countries*, May; and Chidamber, S. R. (2003). An analysis of Vietnam's ICT and software services sector. *Electronic Journal of Information Systems in Developing Countries*, 13(9), 1–11.
31. Gallaugher, J. & Stoller, S. (2004). Software outsourcing in Vietnam: a case study of a locally operating pioneer. *Electronic Journal of Information Systems in Developing Countries*, 17(1), 1–18.
32. *Vietnam Economic News* (2004). *IT Revenue Exceeds Goal*, July 28.
33. Data are from: Tjia, P. (2003). The software industry in Bangladesh and its links to The Netherlands. *Electronic Journal of Information Systems in Developing Countries*, 13(5), 1–8.
34. The Bangladeshi industry association is BASIS (the Bangladesh Association of Software and Information Services). Several of its members are certified to the ISO 9001 quality assurance standard and a handful obtained a CMM-Level 3 certification.
35. Data on Costa Rica are based in part on Mora, A. (2004). Costa Rica/Grupo TecApro, presented at the *IIPI Conference on Strategies for Building Software Industries in Developing Countries*, May.

Chapter 5

1. Inspired by the outsourcing definition in Lee, J. N., Miranda, S. & Kim, Y. (2004). IT outsourcing strategies: universalistic, contingency, and configurational explanations of success. *Information Systems Research*.

2. Wipro, one of India's largest firms, derives roughly one-third of its revenues from contract R&D work.

3. Talent is also tapped through foreign acquisitions. This is practiced by global technology firms that buy technology companies in Israel. Dozens of Israeli firms, from start-ups to established companies, were acquired by foreign technology companies in the period from the mid 1990s through the early 2000s.

4. *Economic Times* (of India) (2003). It's patent mania for US companies in local units, December 17.

5. Granstrand, O., Patel, P. & Pavitt, K. (1997). Multi-technology corporations: why they have distributed rather than distinctive competencies. *California Management Review*, Summer, 39(4).

6. Strassman, P. A. (2004). Most outsourcing is still for losers. *Computerworld*, February 2.

7. Hiskisson R. E., Hitt, M. A. & Ireland, R. D. (2004). *Competing for Advantage*. Ohio: Thomson/Southwestern.

8. Laudon, K. C. & Laudon, J. (2003). *Management Information Systems*. Pearson.

9. *Sand Hill Group* (2003). *The Roadmap to Offshore Success*. Corporate Report, August.

10. Tidd, J. & Trewhella, M. (1997). Organizational and technological antecedents for knowledge creation and learning. *R&D Management*, 27, 359–375.

11. There are two other strategic approaches that need to be mentioned, though they are not unique to offshoring. The *shared services* model is targeted at large companies that wish to bundle internal processes that are common horizontally across divisions. The combined unit may even have P&L responsibility. The clients may choose to create the shared services center offshore, frequently in India. The *co-sourcing* model is a form of joint venture in which the client and provider co-manage a service center.

12. The estimate was made in the mid 1990s and appears in: Narula, R. (2001). Choosing between internal and non-internal R&D activities: some technological and economic factors. *Technology Analysis and Strategic Management*, 13(3).

13. These providers are sometimes referred to as "preferred vendors" and are awarded first opportunities to bid on new projects.

14. *Hindu Business Line* (2002). *India's Treasure Is Its Intellectual Capital*, 7 October.

15. The GE case study is based on a number of internal sources as well public sources including: Bayman, S. (2002). Remarks at the *US–India Business Council Annual Meeting*, June 17. Kripalani, M. & Engardio, P. (2003). The rise of India, *BusinessWeek*, December 8. Waters, R. (2003). It has 3.3 billion USD to buy IT and GE still wants more for less. *Financial Times*, April 30.

16. *Business Standard* (2002). A backup base: GE taps critical mass with its new research lab in Bangalore. June 8.

17. Shared services is a bundling of internal horizontal processes into one entity which may even have P&L responsibility. Some corporations, like GE, choose to create the shared services center offshore.

18. Mishra, P. (2002). GE changes outsourcing paradigm for India. *Express Computer*, April 15.

19. McDougall, P. (2004). GE expects to send more work offshore, May 17.

20. Sengupta, S., Gupta, I. & Singh, S. (2004). GECIS: the house that Jack built … and Jeffrey is about to sell. *Businessworld*, October 11.

Chapter 6

1. The author would like to acknowledge and thank the following colleagues with Mayer, Brown, Rowe & Maw LLP who provided assistance and/or contributed to portions of this chapter. They include Brad Peterson, Paul Roy, Dan Masur, Sonia Baldia, Julian Roskill and Andrew Scott.

Chapter 7

1. Adapted from interview with Prof. W. Chung that appeared in Knowledge@Wharton "Companies that expand abroad: knowledge seekers versus conquerors." March 24, 2004.

2. In essence, knowledge transfer is about going up the "experience curve," which is common for any type of international business activity in which there is transfer of knowledge from one unit to another abroad.

3. A case study of knowledge transfer appears in Nicholson, B. & Sahay, S. (Forthcoming). Embedded knowledge and offshore software development. *Information and Organization.*

4. The discussion of explicit and tacit knowledge is based in part on the following sources: Desouza, K. C. (2003). Facilitating tacit knowledge exchange. *Communications of the Association for Computing Machinery*, 46(6), 85–88. Bassellier, G., Reich, B. H. & Benbasat, I. (2001). Information technology competence of business managers: a definition and research model. *Journal of Management Information Systems*, 17(4), 159–182. Roberts, J. (2000). From know-how to show-how? Questioning the role of ICT in Knowledge Transfer. *Technology Analysis and Strategic Management*, 12(4).

5. Aberdeen Group (2003). Knowledge transfer and on-site/offshore coordination are key to the success of Transco's application maintenance and support program, http://www. aberdeen.com/ 2001/research/090318539.asp

6. IBM's training fund for its employees threatened by offshoring was announced in March 2004.

7. The discussion of governance is based in part on the following sources: Cullen, S. & Willcocks, L. P. (2003). *Intelligent IT Outsourcing: Eight Building Blocks for Success.* Amsterdam: Elsevier. Brown, C. & Magill, S. (1994). Alignment of the IS function with the enterprise: towards a model of antecedents. *MIS Quarterly*, 18(4), 371–403. Sambamurthy, V. & Zmud, R. (1999). Arrangements for information technology governance: a theory of multiple contingencies. *MIS Quarterly*, 23(2), 261–290.

8. Some of the Indian providers, who do not have global coverage, position their Global Office in India rather than near the client's location. Of course, this distance may introduce some difficulties in the relationship between the client and provider. The implication is that the local customer interfaces of these providers take over parts of the communication with the client's global offices.

Chapter 8

1. The experienced reader will note that many programmers like to work alone, without distractions, and usually prefer quiet offices to noisy common work areas. Some programmers will go home to get their work done. This also raises the issue of introversion versus extraversion. But, even most introverted programmers need some social proximity to their colleagues.

2. In fact, when dealing with proximity, other senses come into play: we human beings also crave touch, babies need touch, massage therapy makes us feel more secure, handshakes make us "connect" to those we meet.

3. Kiesler, S. & Cummings, J. N. (2002). What do we know about proximity and distance in work groups? A legacy of research. In: Hinds, P. & Kiesler, S. (eds), *Distributed Work.* Cambridge: MIT Press.

4. Allen, T. (1977). *Managing the Flow of Technology.* Cambridge, MA: MIT Press.

5. Adapted from: Carmel, E. (1999). *Global Software Teams.* Prentice Hall.

6. Kock, N. (2004). The psycho-biological model: toward a new theory of computer-mediated communication based on Darwinian evolution, *Organization Science*.

7. This is what media richness theory, introduced later in the chapter, predicts. Some tasks are not well defined. These are equivocal tasks. Dennis, A. R. & Kinney, S. T. (1998). Testing media richness theory in the new media: the effects of cues, feedback, and task equivocality. *Information Systems Research*, 9(3), 256–274.

8. Various theorists have called this coordination "by mutual adjustment" *cf.* Mintzberg, H. (1993). *Structure in Fives: Designing Effective Organizations.* Englewood Cliffs, NJ: Prentice Hall.

9. Espinosa, J. A., Kraut, R. E., Lerch, F. J., Slaughter, S. A., Herbsleb, J. D. & Mockus, A. (2001). Shared mental models and coordination in large-scale, distributed software development. *International Conference in Information Systems*, New Orleans, LA.

10. The experienced reader will note that we do not always learn to like, love, and trust those whom we work with, as the expression "familiarity breeds contempt" captures so viciously.

11. Hofstede, G. (1991). *Cultures and Organizations: Software of the Mind.* London: McGraw Hill.

12. Herbsleb, J. (2003). Keynote presentation. *The International Workshop in Global Software Development, International Conference on Software Engineering*, 2003.

13. One of the most important formalisms, or structured approaches, is not covered in this chapter. It is the use of formal development methodologies and processes such as CMM (which was introduced in Chapter 1). These formal processes explain, in part, the tremendous success of the offshore industry in India. Formal approaches need to be internally balanced: too much formal software work turns work into a tedious bureaucracy that saps morale. For more on structured approaches, *cf*, Mark, G. (2002). Conventions for coordinating electronic distributed work: a longitudinal study of groupware use, in Hinds and Kiesler (2002). *Ibid.*

14. Maznevski, M. L. & Chudoba, K. M. (2000). Bridging space over time: global virtual team dynamics and effectiveness. *Organization Science*, 11(15), 473–492.

15. A Work Breakdown Structure is a selective outline of the project detailing a hierarchy of tasks and subtasks.

16. One example of a study measuring distractions: Jackson, T. W., Dawson, R. & Wilson, S. (2003). Understanding e-mail interaction increases organizational productivity. *Communications of the ACM*, August.

17. Radicati Group conducted a study that found that work-related instant messaging has grown to roughly 50 billion per year, soon to exceed e-mail in volume. Described in: Kirkpatrick, K. E. (2004). Taking advantage of IM, *Forbes.com*, April 9.

18. Some suggest banning e-mail for content distribution, Majchrzak, A. & Malhotra, A. (2003). *Deploying Far Flung Teams: A Guidebook for Managers.* Society for Information Management, May. Intel has compiled an entire training program to help its employees better manage their e-mail. *Your Time: E-mail Effectiveness Program.* Available on ITshareNet.org

19. A soft goal is that only 10% of the information is pushed over e-mail, while 90% of information is pulled from various repositories.

20. Persistent IM has drawbacks: when people know that their conversations can be tracked they are less candid.

21. Awareness is related to the notions of *shared mental models* and *common ground*. Both of these have been of interest to researchers as they struggle to understand how groups work more effectively, whether they be co-located or distributed.

22. On awareness types also see: Steinfield, C., Jang, C. & Pfaff, B. (1999). Supporting virtual team collaboration. *Proceedings of Group 99.* Available from the ACM. As well as Endsley, M. R. (1995). Toward a theory of situation awareness in dynamic systems. *Human Factors*, (37)1.

23. Cramton, C. (2003). Finding common ground in dispersed collaboration. *Organizational Dynamics*, 30(4), 356–367.

24. Herbsleb, J. (2003). *Ibid.*

25. Jarvenpaa, S. & Leidner, D. (1999). Communication and trust in global virtual teams. *Organization Science*, Winter, 791–815.

26. Travel between distant sites has been labeled "Synchronization By Flying Around."

27. Herbsleb, J. (2003). *Ibid.*

28. Knoll, K. & Jarvenpaa, S. L. (1995). Learning to work in distributed global teams. *Proceedings of the 28th Annual Hawaii International Conference on Systems Science*, pp. 92–101.

29. Morris, M. W. (1999). *E-Mail and the Schmooze Factor*, Research Summary. http://www.gsb.stanford.edu/research/reports/1999/morris.html

30. Scacchi, W. (2002). Understanding the requirements for developing open source software systems. *IEE-Software Proceedings*, February. Scacchi's research was specific to OSS requirements but applies in part to other life cycle phases.

31. Ambler, S. W. (2002). Bridging the distance. *Software Development*, September, www.sdmagazine.com

32. Quote is from a study described in: van Fenema, P. C. & Qureshi, S. (2004). A phenomenological exploration of adaptation in a polycontextual work environment. *Proceedings of Hawaiian International Conference on Systems Sciences*.

33. Espinosa, A. & Carmel, E. (Forthcoming). The impact of time separation on coordination in global software teams: a conceptual foundation. *Journal of Software Process Improvement and Practice*.

34. Adapted from Espinosa, A. & Carmel, E. (2004). *Ibid.*

35. Espinosa, J. A., Cummings, J. N., Wilson, J. M. & Pearce, B. M. (2003). Team boundary issues across multiple global firms. *Journal of Management Information Systems*, 19(4), 157–190.

36. Grinter, R. E., Herbsleb J. D. & Perry, D. E. (1999). The geography of coordination: dealing with distance in R&D work. *International ACM SIGGROUP Conference on Supporting Group Work (Group 99)*, Phoenix, Arizona, ACM Press.

37. Daft, R. L. & Lengel, R. H. (1986). Organizational information requirements, media richness and structural design. *Management Science*, 32(5), 554–571.

38. Even a promising technology like *application sharing* is somewhat limited because there is only one "app" open at a time. Note that when we work face-to-face there is also pointing, tone of voice, and other non-textual messages.

39. There are many who study the impacts of new software tools on software productivity. For example, Software Engineering researchers meet yearly at the Global Software Development workshop that is part of the International Conference on Software Engineering.

40. Knowledge management (KM) systems are now in practice at the Tier-1 IT providers. For example, by the early 2000s Indian-based provider Infosys introduced a corporate-wide KM system with incentives called knowledge currency (KC) to encourage inputs by its software engineers. Some projects even had a designated KM coordinator.

41. Partially based on Hinds, P. & Weisband, S. (2003). Knowledge sharing and shared understanding in virtual teams. In: Gibson, C. B & Cohen, S. G. (eds), *Virtual Teams that Work*, San Francisco: Jossey-Bass.

42. The 2003 study was repeated in 2004. The first study appeared in: Lu, M., Wynn, E., Watson-Manheim, B. & Chudoba, K. (2003). Understanding virtuality in a global organization: toward a virtuality index. *24th International Conference in Information Systems*.

43. 2004 Intel data suggests that this number is underestimating the scale of coordination.

44. In this vein see: Resnick, P. (2002). Beyond bowling together: socio-technical capital. In: Carroll, J. M. (ed.), *Human–Computer Interaction in the New Millennium*. Boston: Addison-Wesley.

45. Best practices for selecting the right people is based on: Majchrzak, A. & Malhotra, A. (2003). *Deploying Far Flung Teams: A Guidebook for Managers*. Society for Information Management, May. And on: Blackburn, R., Furst, S. & Rosen, B. (2003). Building a winning virtual team. In: Gibson, C. B. & Cohen, S. G. (eds), *Virtual Teams That Work*, San Francisco: Jossey-Bass.

46. A useful set of tips for choosing another type of loner – teleworkers (telecommuters) was compiled at Cisco. Cisco is one of the most distributed American technology companies and has an active telecommuting program. See: Smart Valley Telecommuting Guide http://www.cisco.com/warp/public/779/smbiz/netsolutions/find/telecommuting/

47. Carmel outlined the selection criteria for the Global Team Manager above and beyond other traditional leadership qualities. These five criteria were represented in the acronym MERIT: Multiculturalist (the ability to switch cultures); Electronic-facilitator (the ability to manage via technology); Recognition promoter (seeking support for the distant sites at headquarters); Internationalist (conversant in the political-economic issues of team sites); and finally, Traveler (an *easy* traveler). Carmel (1999), *Ibid*.

48. Herbsleb, J. D. & Mockus, A. (2003). An empirical study of speed and communication in globally-distributed software development. *IEEE Transactions on Software Engineering*, 29(3), 1–14.

49. This discussion of dependencies is but a derivation of Dave Parnas' classic essay on modularity and cohesion in software programs. Parnas, D. L. (1972). On the criteria to be used in decomposing systems into modules. *Communications of the ACM*. 15(12).

50. Conways Law appears in: Conway, M. E. (1968). How do committees invent? *Datamation*, 14(4).

51. Architectures for distributing work in software development is based on: Grinter, R. E., Herbsleb, J. D. & Perry, D. E. (1999). The geography of coordination: dealing with distance in R&D work. *Proceedings of International ACM SIGGROUP Conference Supporting Group Work*, pp. 306–315.

52. Krishnamurthy, S. (2002). Cave of community? An empirical examination of 100 mature open source projects. *First Monday*, 7(6), June. www.firstmonday.org/issues/issue7_6/

Chapter 9

1. The study appears in: Swigger, K., Alpaslan, F., Brazile, R. & Monticino, M. (2004). Effects of culture on computer-supported international collaborations. *International Journal of Human–Computer Studies*, 60, 365–380.

2. Hofstede, G. (2003). Lecture at American University, Washington D.C. Author's notes, July.

3. Hofstede's seminal study should be particularly validating to readers of this book, because his data were derived from a very large sample of IBM employees in offices around the world. Thus, in spite of all of them being both members of the computer culture as well as members of the IBM organizational culture, the individuals could still be grouped across major national cultural orientations. Hofstede, G. (1993). Cultural constraints in management theories. *Academy of Management Executive*, 7(1), 81–93. Hofstede, G. (1991). *Cultures and Organizations: Software of the Mind*. London: McGraw Hill.

4. Our recommendations for practical and enjoyable books on cultural differences: Nisbett, R. E. (2003). *The Geography of Thought: How Asians and Westerners Think Differently and Why*. NY: The Free Press. Schneider, S. C. & Barsoux, J. *Managing Across Cultures*. London: Prentice Hall 1997. Hampden-Turner, C. & Trompenaars, A. (1993). *The seven cultures of capitalism*. New York: Currency Doubleday.

5. Yet another implication of the relationship orientation is the need to save face which is very strong in most collectivist cultures. Be careful criticizing people in a group; generally this should be done individually.

6. Hampden-Turner, C. & Trompenaars, A. (1993). *The Seven Cultures of Capitalism*. New York: Currency Doubleday. Trompenaars, F. & Woolliams, P. (2004). *Business Across Cultures*. Capstone.

7. Nisbett, R. E. (2003). *The Geography of Thought: How Asians and Westerners Think Differently and Why*. NY: The Free Press.

8. The table is adapted from the one appearing in: Bennett, M. (1998). Intercultural communication: a current perspective. In: Bennett, M. (ed.), *Basic Concepts of Intercultural Communication: Selected Readings*; Intercultural Press.

9. Constantine, L. (1995). *Constantine on Peopleware*. Englewood Cliffs, NJ: Yourdon Press.

10. This section benefited from two sources: Shewell, C. (2000). *Good Business Communication Across Cultures*. Briston, UK: Mastek; And from presentations and discussions with Lu Ellen Schafer who authored a case later in this chapter.

11. The humorous table "What the English really mean" has been in popular usage by culture trainers for decades. We were not able to determine the original author.

12. Johansson, C., Dittrich, Y., & Juustila, A. (1999). Software engineering across boundaries – student project in distributed collaboration. Working Paper, University of Karlskrona, Sweden.

13. Lu Ellen Schafer is the author of the case study at the end of this chapter.

14. Massey, A. P., Montoya-Weiss, M., Hung, C. & Ramesh, V. (2001). Global virtual teams: cultural perceptions of task-technology fit. *Communications of the ACM*, 44(12), 83–84.

15. Kiel, L. K. (2003). Experiences in distributed development: a case study. *Proceedings of the International Workshop in Global Software Development, International Conference on Software Engineering*.

16. Peters, P. & den Dulk, L. (2003). Cross cultural differences in managers' support for home-based telework: a theoretical elaboration. *International Journal of Cross Cultural Management*, 3(3), 329.

17. Intercultural Press (Interculturalpress.com) has an excellent selection of books on communicating with specific cultures, such as "Encountering the Chinese," "From Da to Yes," (for communicating with Russians), and "American Interactions with Israelis."

18. Edward T. Hall, author of *The Silent Language* (1959) and *The Hidden Dimension* (1969), identified two classic dimensions of culture. Based on his experience in the Foreign Service, his high and low concept refers to the way information is transmitted, or communicated.

19. At the end of the 1990s, Baan got into serious problems and was first acquired by British based Invensys, and later by US-based SSA Global.

Chapter 10

1. Demand for offshore services will continue growing by roughly 20% a year. Meta Group (2004). *METAspectrum Report on the Offshore Outsourcing Market*, October.

2. Lanvin, B. & Quian, C. Z. (2004). Poverty e-readication: using ICT to meet millennium development goals. In Dutta, S., Lancin, B. & Paua, F. (ed.) *The Global Information Technology Report 2003–2004*. Oxford University Press.

3. Kraemer, K. & Dedrick, J. (1999). National policies for the information age: IT and Economic Development. *Center for Research on Information Technology and Organizations*. University of California. http://www.crito.uci.edu/itr/publications/pdf/natl-policiesio-99.pdf

4. Nollen, S. (2004). Intellectual property in the Indian software industry: past role and future need. Paper distributed at the *IIPI Conference Strategies for Building Software Industries in Developing Countries*, Hawaii, May.

5. There are Indian firms successfully exporting software products. They include i-flex solutions and RiteChoice Technologies (financial packages), Eastern Software Systems (ERP software) and Sasken Communication Technologies (telecom products).

6. The discussion of Sri Lanka is based on input from Raja Mitra.

7. Voice over IP, which routes calls over the Internet rather than over traditional telephone circuits.

8. The section on benefits from software exports is based in part on: Carmel, E. (2003). The new software exporting nations: impacts on national well being resulting from their software exporting industries. *Electronic Journal on Information Systems in Developing Countries*, 13(3), 1–6. And on Arora, A. & Athreye, S. (2001). The software industry and india's economic development. *United Nations University, Wider Discussion Paper*.

9. Job creation is also significant in the IT-enabled services (ITES) sector, which requires educated workers in export-focused knowledge services. In 2004, the Indian sector has 245,000 workers, such as in call centers and administrative functions. This sector is also growing fast in other countries. In 2003, there were more than 400 call centers in South Africa, offering employment to almost to 80,000 people (including black Africans). It is estimated that the number of jobs work will increase by more than 200% until 2007. UNCTAD, *World Investment Report 2004. The Shift Towards Services*. United Nations, New York and Geneva, 2004. The ITES sector creates opportunities to young women who would have remained unemployed or would have settled for a less lucrative profession. In Indian call centers, the proportion of women is estimated at 38–68 percent. A job in this sector gives them new confidence and social empowerment, as has not been experienced ever before. It assures a woman, in her twenties, a quality of working life that is much better than what she could have had in traditional feminized occupations. Mitter, S. (2004). *Offshore Outsourcing of Information Processing Work and Economic Empowerment of Women, Presentation at the World Bank*, Washington, June 2.

10. Behrens, A. (2003). Brazilian software: the quest for an export-oriented business strategy.

11. Heeks, R. (1996). *India's Software Industry. State Policy, Liberalisation and Industrial Development*. Sage Publications.

12. Parthasarathi, A. & Joseph, K. J. (2004). Innovation under export orientation. In: D'Costa, A. P. & Sridharan, E. (ed.), *India in the Global Software Industry*.

13. Net earnings from exports are lower than the gross foreign exchange earnings. This is due to expenses related to international travel, living allowances of software workers who undertake their contracts overseas, foreign marketing, multinational profit repatriation, and importation of hardware and software. Indian net earnings are estimated to be around 55% of the gross figures. Joseph, K.J. (2002). Growth of ICT and ICT for Development. Realities of the Myths of the Indian Experience. *United Nations University*, WIDER Discussion Paper No. 2002/78.

14. Carmel, E. (2003). The new software exporting nations: success factors. *Electronic Journal on Information Systems in Developing Countries*, 13(4), 1–12.

15. We often see too optimistic exports targets drafted in national IT strategies. The first REACH initiative of 1999, the strategy for Jordanian ICT development, stated the goal of IT exports of 550 million USD by 2004. This goal has been revised to 100 million USD by 2006.

16. Parthasarathi, A. & Joseph, K. J. (2004). *Ibid. Supra* note 12.

17. Reichgelt, H. (2000). Software engineering services for export and small developing countries. *Information Technology for Development*, 9(2), 77–90.

18. NASSCOM (2004). www.nasscom.org

19. Liu, X. (2004). Technology policy, human resource and chinese software industry. *Proceedings of IIPI Strategies for Building Software Industries in Developing Countries*. Hawaii, May.

20. Terdiman, R. (2001). *CIO Update: A World of Choices for Application Outsourcing*. Gartner InSide, December 5.

21. Nicholson, B. & Sahay, S. (2003). Building Iran's software industry: an assessment of plans and prospects. *Electronic Journal on Information Systems in Developing Countries*, 13(6), 1–19.

22. Behrens, A. (2003). Brazilian software: the quest for an export-oriented business strategy. In: Commander, S. (ed.), *The Origins and Dynamics of the Software Industry in Emerging Market.*

23. Bruell, N. (2003). Exporting software from Indonesia. *Electronic Journal on Information Systems in Developing Countries*, 13(7), 1–9.

24. Nguyen, T. D. (2004). Software industry development in Vietnam. Paper distributed at the *IIPI Conference Strategies for Building Software Industries in Developing Countries*, Hawaii, May.

25. Tjia, P. (2003). The software industry in Bangladesh and its links to The Netherlands. *Electronic Journal on Information Systems in Developing Countries*, 13(5), 1–8.

26. The assumptions used to tabulate the number of IT graduates per year vary quite a bit. These numbers should be viewed with caution.

27. Dijk M. P. van & Wang, Q. (2003). The development of a software cluster in Nanjing. Paper for the EADI workshop on October 30–31, Novara, Italy.

28. Kripalani, M. & Engardio, P. (2003). The Rise of India, *BusinessWeek*, December 8.

29. Indians and Chinese started one-third of technology companies in Silicon Valley between 1995 and 2000. Based on research of Saxenian, quoted in: Immigrant Entrepreneurs and the Bay Area Economy: How Human Capital from Asia Places the Bay Area at the Heart of New Global Networks. Bay Area Economic Forum, Panel Discussion, Spring 2003.

30. Florida, R. (2002). *The Rise of the Creative Class: And How its Transforming Work, Leisure, Community and Everyday Life*. New York: Basic Books.

31. The Business Software Alliance estimates the piracy rates in 2003 in Vietnam, China, and Indonesia at around 90%.

32. The internal use of IT in India is still limited. According to NASSCOM, the Indian software and services exports over 2003–2004 were 12.5 billion USD; the domestic market was only 3.4 billion USD.

33. Behrens, A. (2003). Brazilian software: the quest for an export-oriented business strategy. In: Commander, S. (ed.), *The Origins and Dynamics of the Software Industry in Emerging Markets*. London.

Chapter 11

1. *The Guardian* (2003). Xansa retreats from the continent. December 5.

2. Wells, J. (1998). *IT Services. A Handbook for Exporters from Developing Countries*. International Trade Centre UNCTAD/WTO, Geneva.

3. Sridharan, E. (2004). Evolving towards innovation? the recent evolution and future trajectory of the Indian software industry. In: D'Costa, A. P. & Sridharan, E. (eds), *India in the Global Software Industry.*

4. Screen design is culturally sensitive. An example is the use of colors. Too many colors, especially those that are too bright, are jarring to the Western eye.

5. The Indian NASSCOM sells reasonably priced market intelligence reports on a large number of countries and regions (e.g. Latin America, US, Australia, and Europe). It also publishes reports

on competing countries (e.g. Russia, Ireland, South Africa, and The Philippines). These reports are useful when considering new markets.

6. Tjia, P. (2004). *Offshore Outsourcing*. GPI Consultancy, Markt Rapport, in Dutch, February.
7. Coward, C. T. (2003). Looking beyond India: factors that shape the global outsourcing decisions of small- and medium-sized companies in America. *Electronic Journal on Information Systems in Developing Countries*, 13(11), 1–2.
8. Nicholson, B. & Sahay, S. (2003). Building Iran's software industry: an assessment of plans and prospects. *Electronic Journal on Information Systems in Developing Countries*, 13(6), 1–19.
9. Behrens, A. (2003). Brazilian software: the quest for an export-oriented business strategy. In: Commander S. (ed.), *The Origins and Dynamics of the Software Industry in Emerging Markets*. London.
10. Major e-marketplaces are Elance, ProjectPool, Rent A Coder, Scriptlance, and Smarterwork.
11. Tjia, P. (1999). *Market Survey – Computer Software and IT Services from Developing Countries*. CBI Rotterdam.
12. Moore, S. & Martorelli, W. (2004). *Indian Offshore Suppliers: The Market Leaders*. Forrester Market Overview, April, 7.
13. See http://www.intellectuk.org/groups/offshore/offshore_code_of_conduct.pdf
14. InfoTech (1992). *International studies of Software and Related Services. India's Software and Services Export Potential and Strategies. The World Bank-funded Report for the Department of Electronics Government of India. Volume I and II*. New Jersey: InfoTech Consulting Inc.
15. The 2004 survey of Transparency International lists many other offshore nations as well. The Nordic countries, New Zealand, and The Netherlands are among the least corrupt.
16. International Trade Centre (2002). *Country Profile: Lithuania*. Geneva.
17. Individual country associations usually strengthen their position by participating in both regional and global associations, such as Asia-Oceania Computing Industry Organization (ASOCIO). The dominant international organization is World Information Technology and Services Association (WITSA), a consortium of 60 IT industry associations from different countries. WITSA activities include promoting policies for industry growth and development, facilitating international trade and investment, sharing beneficial knowledge and experience, creating a worldwide contact network, and hosting specific world ICT events.
18. *Elevator pitches* are presentations to make others interested in your services. They are very short, as in the 30 seconds duration in an accidental encounter in an elevator.

Chapter 12

1. McCarthy, J. C. (2002). 3.3 million US service jobs to go offshore. *Forrester News Brief*, November 11, www.forrester.com
2. Forecast by The Australian Computer Society 2003.
3. Hira, R. (2003). On the offshoring of high-skilled jobs. *Testimony to the US House of Representatives Committee on Small Business*, October 20.
4. *Economic Policy Institute, Offshoring, Frequently Asked Questions*. Accessed June 2004. www.epinet.org/content.cfm/issueguide_offshoring_faq
5. Lieberman, J. (2004). *Offshore Outsourcing and America's Competitive Edge: Losing Out in the High Technology R&D and Services Sectors*. Office of US Senator Lieberman, May 11.
6. Deloitte Research (2004). *Making the Off-shore Call: The Road Map for Communications Operators*. Research Report.

7. The happy ending: The hiring firm was very pleased with the four programmers hired and quickly raised the salaries of some. Gumpert, D. E. (2003). US programmes at overseas salaries. *BusinessWeek Online*, December 2.

8. Foote Partners data as appears in: LaFave, R. (2004). Career watch. *Computerworld*, July 5, 2004.

9. By comparison, German labor markets are less flexible due to the cumbersome labor laws. Only 40% are re-employed according the *Economist* (2004). Offshoring: more gain than pain. July 17.

10. US Department of Labor, Bureau of Labor Statistics (2004). Extended mass layoffs associated with domestics and overseas relocations, first quarter 2004. *News Release*, June 10.

11. Tjia, P. (2004). *Offshore Outsourcing*. GPI Consultancy, Markt Rapport, in Dutch, February.

12. Agrawal, V. & Farrell, D. (2004). Who wins in offshoring. *McKinsey Quarterly*, Number 4.

13. United Nations Conference on Trade and Development (2004). *World Investment Report 2004: The Shift to Services*. www.unctad.org

14. Cooper, C. (2004). Poll shows support for offshoring tax. *CNET News.com*, May 4.

15. Computing Research Association data as appears in: Kessler, M. (2004). Fewer college students choose computer majors. *USA Today*, August 8.

16. Lohr, S. (2004). Microsoft, amid dwindling interest, talks up computing as a career. *New York Times*, March 1.

17. Karnitschnig, M. (2004). Vaunted German engineers face competition from China. *New York Times*, July 15.

18. AeA *Offshore Outsourcing in an Increasingly Competitive and Rapidly Changing World: A High-Tech Perspective*, Report, 2004.

19. Lieberman (2004). *Ibid*.

20. Lieberman, J. (2004). *Lieberman Calls Offshore Outsourcing of US Jobs Tip of Economic Iceberg*. Press Release, US Senate, May 11.

21. Ricciuti, M., Frauenheim & Yamamoto. (2004). The next battlefields of advanced technology, *CNET News.com*, May 7.

22. Bonvillian, W. (2004). *Offshoring Policy Options*. Speech given at National Academies. STEP Board, February 20.

INDEX